ELECTRICAL ENGINEERING DEVELOPMENTS

AN INTRODUCTION TO ELECTROSTATIC MEASUREMENTS

ELECTRICAL ENGINEERING DEVELOPMENTS

Additional books in this series can be found on Nova's website at:

https://www.novapublishers.com/catalog/index.php?cPath=23_29&seriesp=Electrical+Engineering+Developments

Additional E-books in this series can be found on Nova's website at:

https://www.novapublishers.com/catalog/index.php?cPath=23_29&seriespe=Electrical+Engineering+Developments

ELECTRICAL ENGINEERING DEVELOPMENTS

AN INTRODUCTION TO ELECTROSTATIC MEASUREMENTS

JOHN CHUBB

Nova
Nova Science Publishers, Inc.
New York

For permission to use material from this book please contact us:
Telephone 631-231-7269; Fax 631-231-8175
Web Site: http://www.novapublishers.com

LIBRARY OF CONGRESS CATALOGING-IN-PUBLICATION DATA

An introduction to electrostatic measurements / John Chubb.
p. cm.
Includes index.
ISBN 978-1-61668-251-4 (hardcover)
1. Electrostatics--Measurement. 2. Electrostatic analyzers. I. Title.
QC589.C48 2009
537'.2--dc22

2010012145

Published by Nova Science Publishers, Inc. † New York

Contents

PREFACE

I have had an involvement with static electricity for over 50 years, on and off. Over the last 25 years or so I have been involved full time in running my own company, John Chubb Instrumentation, whose business has been electrostatic measurements – instruments as well as test methods and some consultancy work. Over the years and a variety of experiences I have developed an appreciation of ways to think about and tackle questions that arise in relation to static electricity. I have presented and published a number of papers on the work I have done and written quite a number of reports for commercial clients. I have also tried to influence the course of events via presentation of papers and drafting of Standards. I have felt that the problems I have faced and the results of trying to deal with questions that arise in electrostatic measurements may well be of interest and help to people who find a need to develop an interest in static electricity as well as those already working in this area. I have therefore tried to bring together my thoughts and experience on electrostatics and to present these in a way that will be helpful. I have not tried to cover areas of electrostatics such as microelectronic machines (mems) and atomic force microscopy or other areas where I have had no direct experience. This is a personal view – the way I have found that matches my appreciation of how the world works and my practical experience. This is not aimed to be an academic treatise or a balanced comparison of work in the field!

My involvement in electrostatics started with the work for my PhD thesis in the Electrical Engineering Department at University of Birmingham. I gained a degree in Physics at Birmingham in 1954. I had expressed the view (in an interview of four final year students for BBC TV) that I did not want to go into research but 'wanted to do something useful'! However, when I was approached by Jim Higham from the Electrical Engineering Department it seemed there was a project that would be both interesting and also useful. The project was to explore the proposal that had been made by Bill Bamford (of British Cast Iron Research Association, at Alvechurch) to Jim Higham that if one took a stream of dusty air, split it into two halves, charged one half positive and the other negative then after combining them the oppositely charged particles would agglomerate and the dust could be collected with usefully enhanced efficiency in a cyclone type dust collector. Once I started work it was very soon evident that it was difficult to electrostatically charge airborne particles without electrostatically precipitating them! It was also evident, in those days before computers, that it was not feasible to do any sensible mathematical modeling of agglomeration. I then turned the work to examining the factors that affected the process of electrostatic precipitation – and much of the PhD thesis was concerned with the approach I developed for observing and analyzing the relative influence of electrostatic and aerodynamic (ionic wind) forces on the behaviour of individual airborne particles in corona discharge fields. Despite the frustrations of research I found I had enjoyed meeting the challenges of developing ways to make observations and measurements in real situations and to analyse and assess the results.

I then spent quite a time away from electrostatics. After a Graduate Apprentiship at English Electric in Stafford I led a small team in the successful development of prototype high power vacuum circuit interrupters able to handle currents of over 10,000A at over 10,000V. This success depended very much on techniques I developed to restrict the area of the contact electrodes available for arcing and using magnetic fields to move the arc roots over the electrode surface to avoid the problem of 'gross melting' at high currents. After moving to the UKAEA Culham Laboratory in 1962 I investigated and developed liquid helium cooled condensation pumping for the high performance high vacuum pumping of large quantities of hydrogen for the controlled fusion studies at Culham. In 1967, and during the following year, there were explosions on a number of the new very large crude oil tankers (VLCCs) during the time they were cleaning their tank by washing with high pressure water jets. These explosions caused massive damage and the loss of one tanker. This was a time for diversification of resources at Culham and Bob Carruthers, head of the Applied Physics Division, suggested that as I had some experience in electrostatics surely I could contribute to solving the problem! This led to quite a major involvement in the international investigation into the causes of the problem. My contribution was development of instrumentation and approaches to observe the electrostatic conditions created in cargo tanks of tankers during tank washing operations and with radio detection, linked to triggered flash photography, to try to understand the conditions associated with the occurrence of spark type electrostatic discharges. These studies were carried out during a number of tanker voyages.

My other main electrostatic project at Culham was development of a novel monitor for airborne fibres – aimed in particular to provide opportunity for portable on-site assessment of asbestos risks. The approach involved use of electrostatic forces to align fibres with scattering of laser illumination to detect and size classify fibres selectively.

After I left Culham in 1978 I worked for a couple of years on computer typesetting at Linotype Paul in Cheltenham as 'Manager, Advanced Technical Planning' (!). After being made redundant (the first of the many!) I took on the job of Managing Director at a small company, IDB, at University College of North Wales, in Bangor. This got me back into electrostatics both for developing and making instruments and for consultancy work. After 2½ years I resigned and set up my own company John Chubb Instrumentation (JCI).

The business of JCI was chosen to be electrostatic measurements. This involved the development of proprietary instruments and methods of testing together with the manufacture and marketing of instruments as well as some consultancy work and testing of materials. A number of contributions have arisen from this work and these will become evident in the body of this book.

My experience in electrostatics, and the other areas in which I have had opportunity to work, have taught me the importance of identifying and focusing on what users and customers are trying to achieve. It has also taught me the importance of retaining an appropriately skeptical view of theory and of received 'wisdom' and the value of learning from experience.

In this book I present the appreciation and experience I have gained from my years of working in electrostatics. I have not attempted to cover every aspect of electrostatics – so this is not an academic textbook. I hope it will prove useful for people with a practical interest in electrostatics and with a need to make and understand measurements.

My work at John Chubb Instrumentation (JCI) over the last 25 years has involved two main aspects – first, the design and development of instruments and appropriate related methods for measurement; and second, consultancy work for a variety of customers. These areas have provided very useful interaction. On one hand I have been able to use JCI instruments to help understand and comment upon customer problems and suggest prospective routes for solution. On the other hand the contact with practical and industrial situations has fed usefully into the design and facilities provided by JCI instruments. There has been good synergy. These two sides have also provided the basis for the useful number of papers I have been able to publish over the years.

Much of the consultancy work I have done has involved a day on-site and a day in preparation of a report. Some studies have been longer, for example the studies on solvent extraction operations in Chile, and some have involved several further testing of samples for a number of customers. One basic problem from my experience in consultancy (and this includes the consultancy work I did while at UKAEA Culham Laboratory) is that rarely does one have much involvement with the outcome or consequences of one's consultancy work. One does not see either the practical problems of implementing one's recommendations or the economic costs or benefits. In this respect it can be difficult to get feedback to benefit future consultancy work. For these reasons I am glad the manufacturing side of the business of JCI was the dominant activity – at about 85%. This is not to belittle the value of consultancy, but to keep it in perspective.

Over the years I feel that useful contributions to the area of electrostatic measurements have been made by, for example:

- development of 'field mill' fieldmeters - in particular those not needing earthing of the rotating chopper
- development of fieldmeters able to make measurements in adverse environmental conditions
- development of instrumentation and methods for assessing the electrostatic suitability of materials by measurement of corona charge decay
- development of the concept of capacitance loading
- methods for measuring the electrostatic shielding of materials and their use to measure to resistivity of materials with buried conductivity
- methods for the calibration of electrostatic measuring instruments.

The main satisfaction from the work over the last 25 years at JCI has been from the large number of JCI instruments in use in many and varied industries around the world. A number of companies evidently rely on JCI measurements on a regular and continuing basis. This shows we must have done something right from the point of view of users! When the merits of the approaches to measurements we have developed come to be more widely recognised it is hoped that these will lead to constructive discussion in committees considering International Standards.

In February 2009 the assets of JCI were sold to Chilworth Technology Ltd in Southampton. It is they who are now responsible for the manufacture and marketing the instruments developed over the years by JCI.

John Chubb - Cheltenham, UK. 2010 (http://www.infostatic.co.uk)

Chapter 1

INTRODUCTION

'Static electricity' arises when materials in contact are separated and the speed of charge movement on one or the other surface is slow compared to the timescale of separation.

If the time for charge movement over surfaces is very short then the charges of opposite polarity can move to the last point of contact and neutralise each other – and then little charge is retained. This occurs when, for example, two metals are separated. If the time for charge movement on one material is very long then charge will be retained on that material when the surfaces separate - it is 'static'. If one of the contacting surfaces is a good conductor, but it is electrically isolated, then although charge can move across its surface the charge will remain trapped on the conductor. With static 'charging' no charges are actually created - only separated.

The 'charge' separation arises from differences in the electronic structure of the two surfaces. It is very much a surface effect and so very much influenced by the materials and by any surface treatment or contamination.

Charge separation occurs at individual surface separation events as well as in multiple and repetitive events. In the later case the repetitive separation is effectively a 'charging current' generator. The difficulty experienced by charge leaking away is then usually thought of, and described, as a 'resistance'. The limitations of this way of thinking will become apparent later.

This book is concerned primarily with static electricity in industry and practical situations. Much of the interest concerns the hazards and problems which static can cause The following list indicates the range of ways that static can be of importance:

- *ignition of flammable gases*: petrochemical, explosives, plastics and printing and pharmaceutical industries (loss of plant/production, damage to personnel)
- *shocks to personnel:* paper, packaging and printing industries (indirect risks of accidents and damage)
- *attraction of dust and debris:* packaging and printing (affects performance and cosmetic appearance)
- *cling of thin films:* packaging film handling, lingerie (cling causes difficulties in handling, in wear and in appearance)
- *damage to semiconductor devices*: (added costs in production and poor reliability)
- *upset operation of microelectronic systems*: relevant to malfunction of computer equipment, instrumentation and process control

Static electricity can also used constructively to transport and hold particles to surfaces - for example: electrostatic precipitation, liquid spraying, electrostatic clamping and electrophotographic copying. Electrostatic forces can also be used for particle alignment – for example, in flocking. Electric fields in the atmosphere also provide information on cloud processes and the conditions for the occurrence of lightning.

It is hoped that this book will inspire confidence:

- in the ability to appreciate how, where and why static arises
- in making meaningful electrostatic measurements and assessing their significance
- in considering actions appropriate to control or to use static

With a better appreciation of static there is a better chance to recognise prospective problems in advance and choose materials and/or adjust processes to avoid risks and problems and optimize constructive use.

It is important to appreciate that electrostatics is only ever a part of the overall 'system'. Electrostatic aspects must be viewed in conjunction with many other factors – such as the people involved, engineering practicality and the economics.

This book starts with an outline of basic electrostatics followed by a discussion on measurements. These topics may seem a bit 'academic' to some people. However they are important because they form the basis for understanding and assessing practical problems – both the electrostatic conditions present in practical situations and the suitability of materials for applications. Not many people think about electrostatics as easily as they think about current flow and Ohm's Law - so it is important to appreciate and come to terms with the basic concepts. In addition to references of published papers, at the end of each main section, a number of sources of information are listed later in the book - such as books, scientific journals, organisations running meetings, useful websites and sources of Standards documents. Some additional specific information is provided in the Annexes that it is hoped will prove useful.

It is hoped this book will be a useful guide and reference source - and will stir the imagination and enthusiasm of readers!

Chapter 2

BASIC ELECTROSTATICS

2.1. INTRODUCTION

Static electricity is associated with materials on which electrical charge moves only slowly (insulating materials) and with electrically isolated conductors. 'Insulation' and 'isolation' prevent easy migration of charge. So charges stay in place - are 'static'. It is the effects that the charges produce that are then important.

The parameter of basic importance in electrostatics is 'charge'. It is *charge* which gives rise to the electric fields generating forces that attract thin films and particles to surfaces and *charge* which gives rise to high voltages in low capacitance systems. High voltages can cause sparks. If these have sufficient energy they may ignite flammable gases, cause shocks to personnel, damage semiconductor devices and upset the operation of microelectronic equipment.

This section introduces basic concepts of electrostatics and provides a framework for understanding and quantitatively assessing electrostatic questions and problems. A good textbook is Feynman 'Lectures on Physics' [1]. Other good sources are available [2,3].

2.2. CHARGE

Static electricity arises as the separation of positive and negative charges at the interface between two dissimilar surfaces. If one or other of the surfaces prevent easy migration of charge, or the conductor on which they reside is isolated, then this charge is 'static' on the surface and remains available to influence the surroundings. 'Static' electricity can also arise on surfaces as trapped ions from the air.

Static charges may be electrons or positive, or negative, ions - but they are in the basic units of electronic charge 1.602×10^{-19} coulomb. On a surface there are some 10^{19} atomic sites per m^2 - so if the charge of even quite a small fraction of the surface atoms is changed then quite large quantities of charge are easily involved and, as we will see, quite large potentials and electrostatic forces can arise.

2.3. Force between Charges

The force between two charges, Q_1 and Q_2, is proportional to the quantities of charge, and inversely proportional to the square of the separation distance, d:

$$F = - Q_1 Q_2 / (4 \pi \varepsilon_0 d^2)$$

This is Coulomb's Law. The force F is in Newtons, the quantities of charge Q_1 and Q_2 are in Coulombs and the separation distance d is in meters. The constant ε_0 is called the 'permittivity of free space' and has a value 8.854×10^{-12} (see later). Charges of the same polarity repel each other and of opposite polarity attract.

Everyday examples of electrostatic forces are the attraction between thin layers of plastic film and of fine dust particles to surfaces.

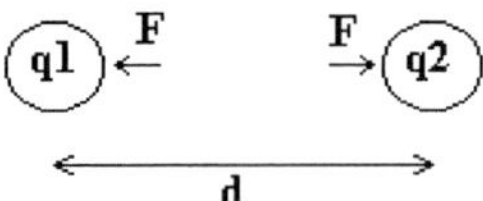

2.4. Electric Field

Around a charged body there is a force of attraction or repulsion for any other charges. This force, for a unit of charge, at any point is called the 'electric field', E, at that point.

The direction of the field at any point depends on the direction of force on a positive charge there - and is hence a vector parameter. Diagrammatically it is useful to draw 'lines of force' to represent the electric field. The closer together the lines of force the stronger the electric field.

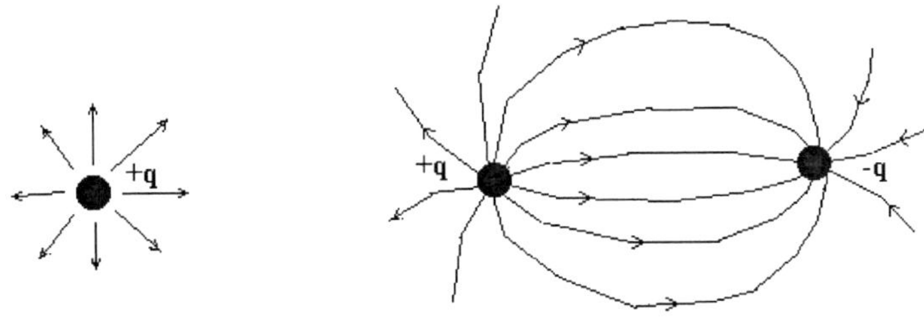

From Coulombs Law the electric field strength E at a distance d from a charge Q will be:

$$E = Q / (4 \pi \varepsilon_o d^2)$$

The constant of proportionality $1 / (4 \pi \varepsilon_o)$ is, for historical reasons, defined in the MKS system of units by the relation with c as the velocity of light in vacuum:

$$1 / (4 \pi \varepsilon_o) = c^2 \times 10^{-7}$$

The constant ε_o hence has a value 8.854×10^{-12} coulomb^2/(Newton m^2) or coulomb / (volt m). This is the 'permittivity of free space' noted above.

2.5. POTENTIAL

The 'potential' at a point is defined as the amount of work needed to bring a unit charge from infinity to that point. The potential difference between two points is then the work done to move a unit charge between these two points. The work done does not depend on the route followed so the potential is a scalar quantity.

The unit of potential is the 'volt'. The potential at a radial distance d from a single point charge Q is:

$$V = Q / (4 \pi \varepsilon_o d)$$

The electric field E ($V m^{-1}$) relates to potential as:

$$E = -\nabla V$$

2.6. Electric Flux Density

The electric field from an isolated point charge is radial. The field may conveniently be represented diagrammatically by lines of force. The 'flow' or flux of electric field across any spherical surface enclosing a point charge Q will then be Q/ε_o.

2.7. Gauss' Law

Gauss' Law states that the nett flux through any surface equals the nett charge enclosed within that surface.

$$\nabla . E = q / \varepsilon_o$$

For example, for a uniform distribution density of charge ρ ($C m^{-3}$) in a sphere of radius r the total charge is $4/3 \pi r^3 \rho$. The outward flux will be $4 \pi r^2 E$. Hence the electric field at the sphere surface is:

$$E = \rho r /(3 \varepsilon_o)$$

The maximum voltage at the centre of such a spherical distribution of charge is:

$$V = \rho r^2 /(6 \varepsilon_o)$$

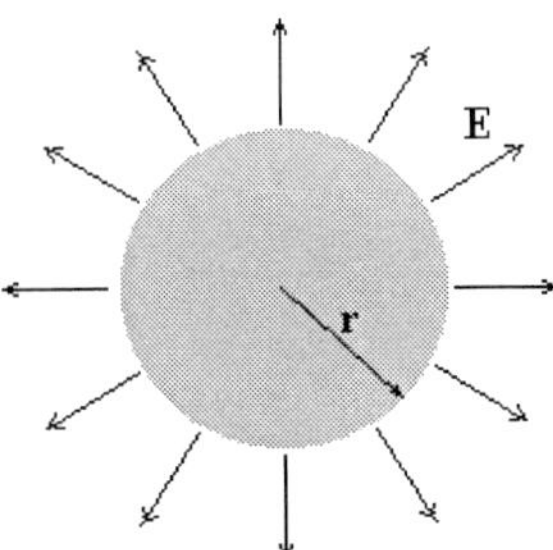

For say a 1m radius spherical distribution of charge density of 10^{-6} C m^{-3} the electric field will be 37.6 kV m^{-1} and the maximum potential at the centre 18.8 kV.

Similarly for a cylindrical distribution of charge density:

$$E = \rho\, r /(2\, \varepsilon_o) \text{ and } V = \rho\, r^2 /(4\, \varepsilon_o)$$

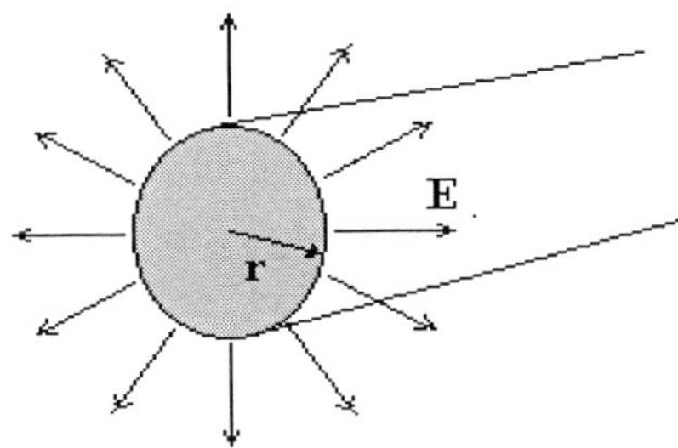

2.8. Induced Charge

The electric field from each side of a sheet of uniform charge density σ (C m^{-2}) is:

$$E = \sigma / (2\, \varepsilon_o)$$

There can be no electric fields inside a conductor and the electric field outside is:

$$E = \sigma / \varepsilon_o$$

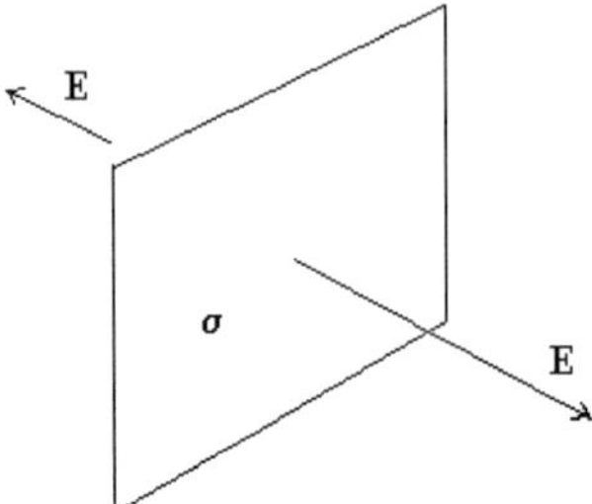

There can be no electric fields inside a conductor and the electric field outside is:

$$E = \sigma / \varepsilon_o$$

Thus a conducting body placed in an electric field will have charge 'induced' on it with the density of charge proportional to the electric field at the surface.

The charge induced on a conductor is of opposite polarity to that of the charge source. Opposite charges attract so there is an attraction between a charge and a conducting surface. For a plane conducting surface this 'image charge' is equivalent to an equal charge at an equal distance behind the surface, so calculation of the force involves a distance equal to twice the separation distance from the conducting surface.

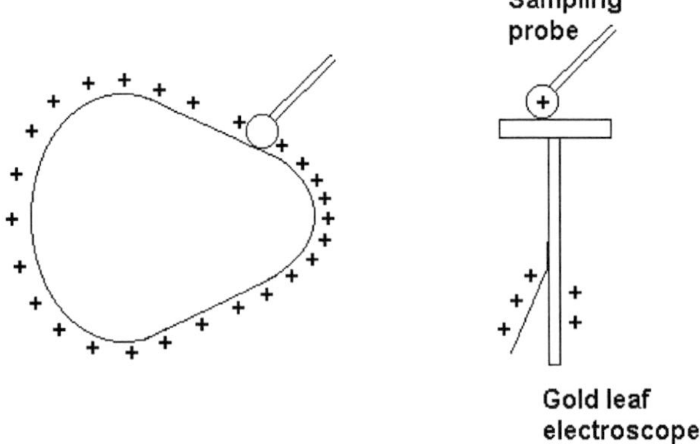

If a small conducting body (e.g. a small sphere) on an insulating handle is touched to a conducting body charge will be shared and the probe will acquire a charge dependent of the local electric field there - related to the local surface charge density. If the amount of charge can be tested then we have a means to check if the conducting body was charged and to explore the density of charge over that body. A 'gold leaf electroscope' indicates charge by repulsion between a fine gold leaf and it's nearby conducting support and the angle of the leaf indicating the level of charge from the level of repulsive force. Sharing charge from a conducting probe with a gold leaf electroscope provides a good classical method to indicate and measure charge. Such observations show that the electric field (and charge density) on a conducting body is inversely proportional to the local radius of curvature.

2.9. Capacitance

For two conducting plates of area A separated by a distance d (m) with a voltage difference between them V the electric field E between the plates is:

$$E = V / d$$

and the quantity of charge on each of the plates:

$$Q = A\,\sigma = \varepsilon_0 A\,E = \varepsilon_0 A\,V / d$$

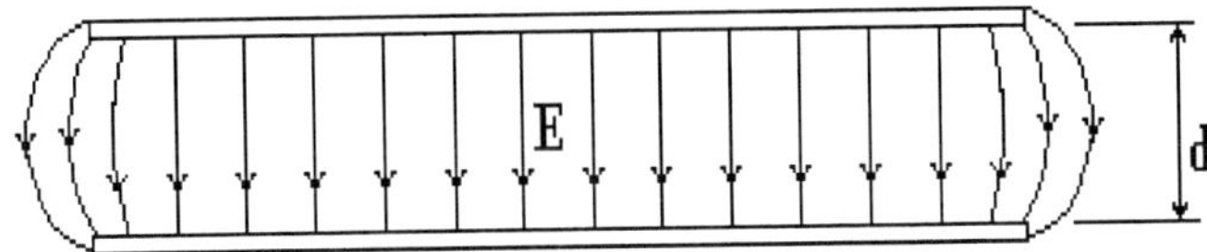

The proportionality between the quantity of charge Q and the voltage V is called the 'capacitance' C of the system. The unit of capacitance is the 'farad' with units of coulomb/volt. For the pair of parallel plates:

$$C = Q/V = \varepsilon_0\,A / d$$

This relation is not quite accurate because the electric field is not uniform near the edges of the plates and extends outwards from them a bit.

For an isolated sphere of radius a (m) the capacitance is $4 \pi \varepsilon_o a$ – hence a sphere of 10mm radius the capacitance is about 1 picofarad (10^{-12} farad). The capacitance for a number of simple geometric arrangements (for instance concentric spheres, concentric cylinders, parallel rods and rods or spheres near plane conductors) can be calculated analytically [2,6]. These can provide useful guidance for estimating capacitance in practical situations. Alternatively capacitance values may be calculated by computer modelling calculations [7,8,9] or, with suitable care, they can be measured.

2.10.ENERGY

The electrostatic energy U (joule) of a capacitor of C (farads) charged to a potential V is:

$$U = 0.5\, C\, V^2$$

2.11. POISSON'S EQUATION

$$\nabla^2 V = -\rho/\varepsilon_o$$

where ρ is the volume density of charge ($C\ m^{-3}$). If no charges are present then Poisson's equation becomes the Laplace's equation:

$$\nabla^2 V = 0$$

There are a number of geometric forms for which the above equations can be solved analytically. Two and three dimensional finite element and finite difference computer modelling programs are available to find potential and electric field distributions within practical geometric arrangements [7,8,9].

2.12. DIELECTRICS

If an insulator is placed between the plates of a parallel plate capacitor the capacitance is increased.

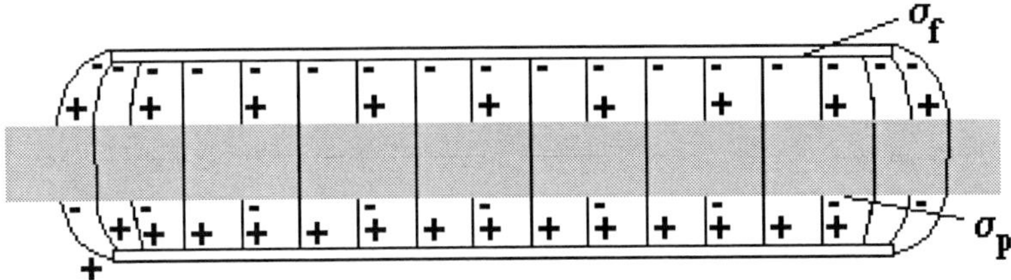

If a slab of conductor were placed between the capacitor plates the field in the remaining gap is increased. A similar effect occurs with insulating materials, but because charges cannot move through or over the surface of insulators the material becomes 'polarised'. The atoms or molecules distort slightly under the electrical stress, polarise, and while these effects cancel out within the volume of the insulator they appear as surface charges on either side of the block of material.

Applying Gauss's Law to the surface S the electric field in the insulator is:

$$E = (\sigma_f - \sigma_p) / \varepsilon_o$$

where σ_f and σ_p are the charge densities on the free surface of the capacitor plates and the surface of the insulator. The polarisation charges are proportional to the electric field and this is usually written:

$$P = X\, \varepsilon_o\, E = \sigma_p$$

$$E = \sigma_f / (\varepsilon_o (1 + X))$$

where X is the 'electric susceptibility' of the insulator/dielectric. This shows the reduction of the electric field within the insulator compared to that just outside.

The total charge on the capacitor plates is $\sigma_f A$ so the capacitance by Q = CV becomes for a capacitor fully filled by an insulator:

$$C = \sigma_f A / V = A\, E\, (\varepsilon_o (1 + X)) / (E\, d)$$

$$C = \varepsilon_o A (1 + X) / d$$

The increase in capacitance for this fully filled capacitor is the factor (1+X) and this is usually called the 'permittivity' of the material - k.

2.13. Charge Migration and Dissipation

Insulating materials have very few free electrical charges so only very small currents flow under the influence of an electric field. In more usual terms the volume 'resistivity' may be from 10^{12} to 10^{16} ohm m and the surface resistivity over 10^{10} ohms per square. Not only are there few charge carriers but their number and the velocity of their movement may depend in a non-linear way upon the electric field. In this situation 'resistivity' is not a very useful parameter for characterising materials. Of more direct practical importance for the control of static is the timescale for charge movement. If the timescale for charge movement is less than the timescale for charge generation then no significant charge levels or surface voltages will arise and there will be no static problems. Conversely, if the timescale is longer then problems from static can be expected.

For uniform materials, such as liquids, it is expected that the time for charge dissipation τ (σ) can be related to the volume resistivity ρ (ohms) and permittivity k as:

$$\tau = k \varepsilon_o \rho$$

As we shall see, in Chapter 4.9, this relationship is not very useful for assessing the electrostatic suitability of solid surfaces and practical materials.

2.14. Charge Separation Mechanisms

Contact between surfaces results in transfer of charge as electrons to match up energy levels of neighbouring atoms around the points of contact. The flow of charge continues until inhibited by the potential difference, the 'contact potential', between the surfaces. On separation the surfaces may retain the transferred charge if the speed of movement is slow.

Charge transfer depends on differences in 'contact potential' and materials can be listed as a 'triboelectric series'. Charge separation depends on surface conditions. It is strongly affected by contamination and by the speed and pressure of rubbing actions. Only in a few instances, for example in photocopiers and in some paint spray situation, are triboelectric effects able to be used to determine charging in a defined and controlled way.

Charge separation also occurs when a liquid meets a conducting surface. Ions formed by dissociation in the fluid or from trace impurities will adsorb at the boundary surface. This creates the so-called 'double layer' at the boundary. An appreciable charge can be adsorbed to the surface - which is matched by charge in the fluid. With relatively insulating fluids the charge in the volume can be swept away by fluid flow as a 'streaming current' which can be in the range 10^{-10} to 10^{-7} A. A double layer also occurs when water breaks away from a surface - and this is responsible for the formation of the highly charged mists that can arise when high pressure water jets hit surfaces.

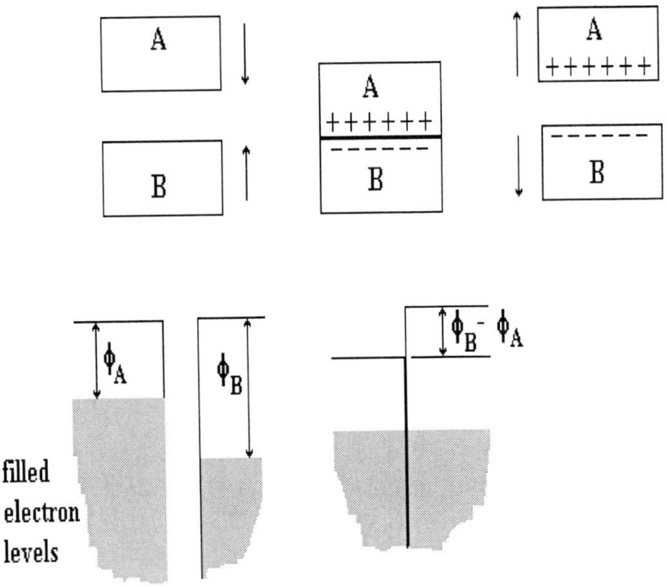

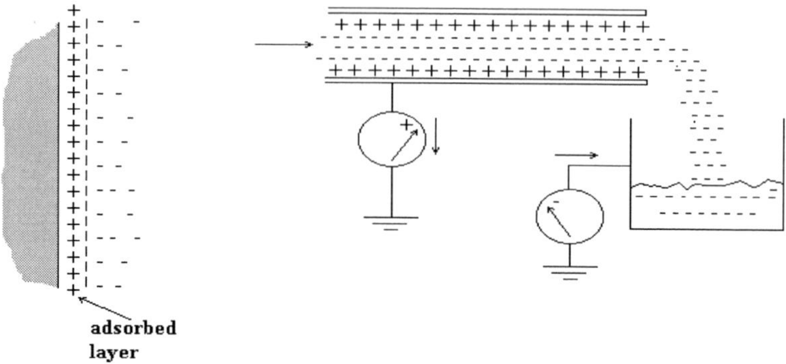

2.15. Electrical Breakdown

If an electric field, in air at normal temperature and pressure, increases above about 3 MV m^{-1} then 'electrical breakdown' occurs. If a loose electron or ion arises it will be accelerated in the local electric field. If it picks up enough energy from the field before it hits another air molecule it may cause ionisation of that molecule. This can lead to further ionisation and rapid growth of the number of ions present. This multiplication of ionisation can destroy the insulating properties of the air so the air becomes conducting and a 'spark' type discharge may occur.

If the electric field exceeds the critical value for ion multiplication in just a localised region, for example near a sharp point, while the average electric field over the gap to nearby conductors is quite low, then a localised discharge takes place with a fairly low current (up to perhaps a mA) between the electrodes. This is a 'corona' discharge. If the electric field exceeds the breakdown strength of air over several mm, for example with an electrode of several mm radius of curvature, then the discharge involves more current and has a condensed luminous channel which yet does not reach to another electrode. This is a 'brush discharge'.

If the breakdown strength of air is exceeded near an electrode and also the average electric field over the remaining gap to the other electrode exceeds about 500 kV m^{-1} then breakdown is likely to occur as a spark discharge between the electrodes. A spark discharge involves currents of many mA or A, is usually of durations from ns to μs and is accompanied by the emission of heat, light and sound.

The voltage for electrical breakdown depends on the geometry of the gap, on the gas and on the pressure. As the pressure or gap is reduced the breakdown voltage decreases to a minimum and then rises - following the Paschen Law curve [2,4]. For air the minimum is about 300 volts. A good textbook and reference book on electrical breakdown of gases is that by Meek and Craggs [4]. For small gaps and low gas pressures electrical breakdown may occur via field emission processes at the electrode surfaces.

If a high electric field arises at the surface of a fluid that is at least partially conducting then the surface will distort and break up [5]. This occurs at lower electric field values than that for air breakdown for continuously applied electric fields.

REFERENCES

[1] R. P. Feynman, R. B. Leighton, M. Sands *'The Feynman Lectures on Physics'* Volume II. Addison-Wesley Publishing Co, USA 1964.

[2] D. M. Taylor, P. E. Secker *"Industrial electrostatics: Fundamentals and measurements"* Research Studies Press, John Wiley 1994

[3] H. Haase, *"Electrostatic Hazards: Their evaluation and control"* Verlag Chemie, Weinheim, New York, 1976

[4] J. M. Meek, J. D. Craggs *'Electrical breakdown of gases'* Oxford, 1953

[5] G. I. Taylor, A. D. McEwan *'The stability of a horizontal fluid interface in a vertical electric field'* J. Fluid Mech 22 1965 p1.

[6] Smythe W R *"Static and dynamic electricity"* 1968 McGraw-Hill, New York

[7] Trowbridge, C. W. *"Computer modelling of electrostatic fields"* Electrostatics 1991, Inst Phys Confr Series 118 p253

[8] Thomas, C. L. *"POTENT A package for the numerical solution of potential problems in general 2D regions"* Proc Confr on Software for Numerical Mathematics and its Applications, Loughborough Univ, April 1973 Academic Press, London

[9] Thomas, C.L. *"THREE-D A digital computer code for the design and analysis of three-dimensional electrostatic fields"* IEE Confr Computer Aided Design, Univ Southampton, April 1974

Chapter 3

MEASURING INSTRUMENTS

3.1. Introduction

Measurements are needed to show you the electrostatic conditions that are present, to help understand why these conditions are present, to help select and develop materials and control methods appropriate for particular situations and to help confirm that control measures put in place are continuing to work. Without measurements it is all guesswork!

The classical groundwork in electrostatics used instruments such as the gold leaf electroscope and the quadrant electrometer. These are perfectly valid measuring instruments - but they require specialist use and really belong to the physics laboratory. The instruments described below are appropriate for use by non-specialists in industrial as well as laboratory situations.

An important area of measurement is assessing the characteristics of materials. Are materials in use, or being considered, suitable for their purpose? And when thinking about 'suitability' this is not only what is relevant for the immediate area of application (for instance within a processing operation) but also through to the final 'end user'.

Guidance is provided in a number of Codes of Practice [1,2,3,4] on arrangements and procedures designed to avoid static problems and risks. However, there are always situations and materials where there is uncertainty. Is there a problem, what is it due to, will suggested remedial actions work, are alternative approaches prospectively more suitable, and are remedial actions continuing to work effectively?

In many cases measurements only need to be informative and to provide guidance. But measurements may have contractual, legal or safety implications. In such situations the instruments used need to be:

- appropriate in capability for the measurements required
- used with appropriate measurement procedures that include recording and interpretation of observations
- formally calibrated

The following sections describe the basic instruments needed for assessing the electrostatic conditions that arise in various situations and the instruments needed to assess the electrostatic characteristics of materials.

The basic instrument for assessing electrostatic conditions is the electrostatic fieldmeter. This can be used in a wide variety of ways. Assessment of the characteristics of materials may involve measurement of charge with a Faraday Pail, with measurement of the ability of surfaces to dissipate static charge, measurement of the shielding characteristics against electric field transients and measurement of the opportunity to create incendiary discharges. The following descriptions draw attention to the instrument design features that are needed to ensure that measured values can be used with confidence. The ways in which these basic instruments may be used in practice are discussed in Chapter 4.

3.2. Measurement of Electric Field

3.2.1. Basic Aspects

The parameter of primary importance in static electricity studies is 'charge'. In some practical situations this can be measured directly, however as it is via the electric fields created by charges that items nearby are influenced. Measurement of electric field hence provides the most generally applicable way to detect, locate and quantify sources of static charge and to assess the influence of such charge.

Measurement of electric field also provides the basis to measure a whole variety of other parameters of interest in electrostatics - voltages on surfaces, on people and in volumes, the density of charge on surfaces and in volumes, the nett quantities of charge on items and the ability of materials to dissipate static charge.

Electric fields may be measured by the mechanical forces created. This has rather low sensitivity and for quantitative measurement requires geometries that define the electric field to be very uniform [5]. The more practical approach is to measure the charge induced on the sensing surface of a 'fieldmeter' instrument.

An electric field is associated with a distribution of charge at a conducting surface - as discussed in Chapter 2.8:

$$E = \sigma / \varepsilon_o$$

Where E is the electric field (V m^{-1}), σ is the area charge density (C m^{-2}) and ε_o is the permittivity of free space (8.854 10^{-12}). This relationship forms the basis for all 'fieldmeter' instruments – instruments to measure electric field.

3.2.2. Induction Probe Type Fieldmeters

The simplest form of fieldmeter is the 'induction probe'. This is illustrated in Figure 3.1. It has an exposed conducting sensing surface on which the electric field to be measured induces a charge with the surface connected to a measurement circuit. For a surface of area A:

$$Q = \sigma A = E A \varepsilon_o$$

The charge held at the surface is matched by an equal charge repelled to the measurement circuit.

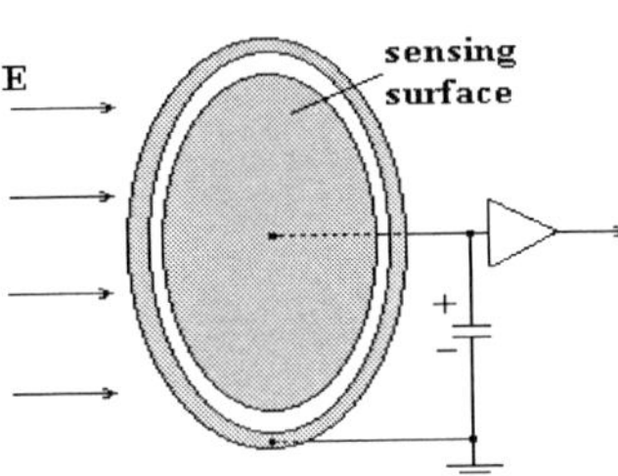

Figure 3.1. Induction probe type fieldmeter.

The quantity of charge induced on the area of the sensing surface can be measured in two main ways:

- from the voltage developed across an input capacitor of a high impedance preamplifier
- using a virtual earth charge measurement circuit

Two basic input stage circuits are shown in Figures 3.2 and 3.3.

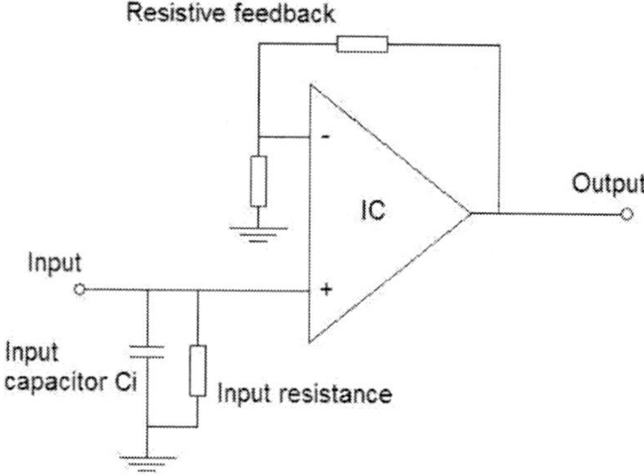

Figure 3.2. Measurement of charge via voltage on a capacitor. (The 'input resistance' may be just the leakage resistance of the circuit connections and amplifier)

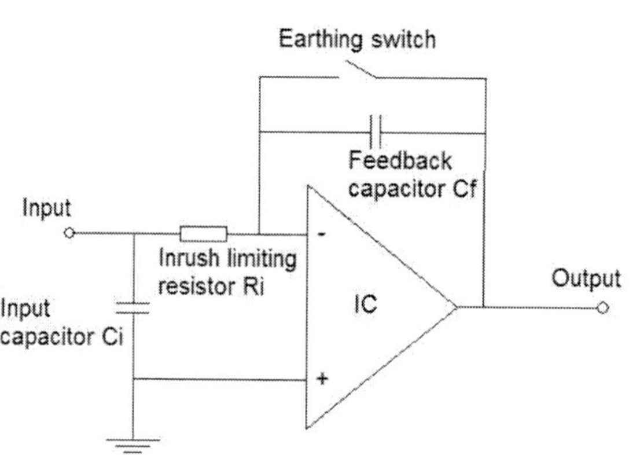

Figure 3.3. 'Virtual earth' charge measurement circuit.

The main limitations of induction probe instruments are that they cannot be used for long term continuous monitoring and that they are they are not very sensitive. If an induction probe is presented with a step function of electric field the output will step in proportion to the input variation of field, but will then decay away in relation to the RC time constant of the input or feedback capacitor and the total leakage resistance to the input connections. There will also be adverse influence from voltage and current offsets of the input amplifier stage. This leakage time constant limits the maximum time over which it is sensible to make measurements and the maximum sensitivity that can therefore be realised. Higher sensitivities can be achieved where short observation times are acceptable.

Induction probe instruments need to be 'zeroed' in a region that is free of electric fields. If this is not done then the output 'zero' will correspond to the level of electric field at which 'zeroing' was performed. For input voltage measurement circuits 'zeroing' is achieved, in zero electric field conditions, by simply shorting the input to the circuit earth. With virtual earth charge measurement circuits it is necessary to short out the feedback capacitor. In both situations the switch needs to avoid reducing leakage resistance.

The voltage measurement circuit approach requires the greater carc in the insulation of the sensing surface and its direct connections because actual voltages are generated on the input connections. With the virtual earth charge measurement circuit the sensing surface is held at earth potential. The requirement for very low leakage hence applies primarily across the feedback capacitor and zeroing switch rather than the mounting of the sensing suface. In both cases it may be useful to use guard rings or make electrical connections via air linkages to avoid leakage on or in the surface of circuit boards.

Where very fast edge electric field transients may need to be handled the voltage measurement circuit is to be preferred. This is because the performance of the virtual earth charge measurement circuit is limited by the output drive current capability of the amplifier to charge the feedback capacitor sufficiently fast to prevent the input terminal going to the power supply rails. This limitation can be overcome, as illustrated in Figure 3.2, by the addition of the 'input capacitor' Ci and the 'inrush limiting resistor' Ri.

Induction probe instruments are not suitable for measurements in the presence of ionised air. This is because unipolar ionised air will create a current flow to the sensing surface trying to neutralise the effect of the electric field and bipolar ionised air will provide an unquantified leakage resistance path between the sensing surface and earth.

3.2.3. 'Field Mill' Fieldmeters

'Field mill' instruments overcome the limitations of simple induction probe instruments by using an earthed chopper to modulate the electric field to be measured and which induces charge at the sensing surface [5,6]. A basic arrangement is shown in Figure 3.4. The charge induced on the sensing surface may be measured using either of the input circuits shown in Figure 3.2 or 3.3. The alternating signal generated at the sensing surface can then be amplified and phase sensitive detected to provide an electrical output that is linearly related to the strength and to the polarity of the incident electric field.

It is equally appropriate to work with a sectored chopper and sectored sensing surface or with a sectored sensing aperture plate, a sectored chopper and a plane sensing surface. Instruments are usually constructed with an axial geometry (as in Fiure 3.4) but instruments may equally well be made with radial geometry [7]. Axial fieldmeters are used (as discussed more fully in Chapter 4) either mounted flush with a large plane conducting to measure the electric field there or used on their own to measure voltages of surfaces nearby or local space potentials in a volume.

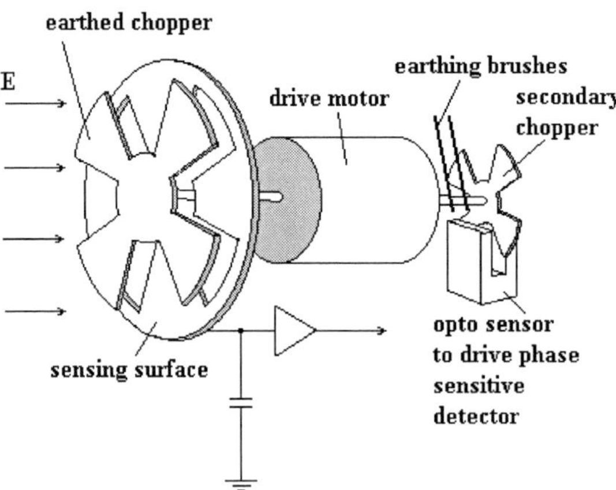

Fig 3.4. Basic 'field mill' fieldmeter.

'Field mill' instruments are mechanically more complex than induction probes but they provide much higher sensitivities, long term stable zeros and avoid the need to 'zero' the instrument in a region free of electric fields each time they are switched on. They are also suitable for use in the presence of ionised air.

The opportunity for much higher sensitivity arises because the RC leakage time constant of the input connection or the charge measurement feedback only needs to be suitably longer than the timescale at which the electric field is modulated. As the modulation time is likely to be 0.05s or less it is easy to see that the basic sensitivity may easily be 1000 times greater than for induction probe instruments.

Measuements may be made in the presence of ionized air because 'conduction' charge signals arise with a 90 degree phase shift from the induction charge signals and so are rejected at phase sensitive detection.

Measurements with field mill type fieldmeters can be made with sensitivities to a few V m^{-1}, with response times down to a few ms and with accuracies to 1%. High useable sensitivities are useful because they enable even small quantities of static charge to be detected with confidence at a distance. High accuracy is not needed in many practical studies such as in the assessment of electrostatic conditions around the workplace. Precision, stability and low noise are however important when measurements are needed of small changes in potentials such as in the measurement of very slow rates of change of electric fields in measurements of charge decay.

Fieldmeters are also available that are based on the use of 'voltage follower' designs where the voltage of an electric field sensing element within an earthed case is adjusted by servo control to null the electric field sensed [8]. The electric field at the instrument sensing aperture is then derived from the nulling voltage applied and dimensions of the sensing arrangement. The main limitation of this approach is that the sensitivity is much lower than for a rotating chopper type fieldmeter because the sensing aperture area is much smaller and because of limitations by the amplitude and area of vibratory movement. Internal gaps are also small, so there is also greater susceptibility to surface contamination.

3.2.4. 'Field Mill' Fieldmeters Without Earthing of the Rotating Chopper

Traditionally 'field mill' fieldmeter instruments have been based, as illustrated in Figure 3.4, on the use of an earthed rotating chopper to modulate the observed electric field at a sensing surface [5,6]. This approach works well, but has a number of limitations for practical and commercial instruments.

Making a good low noise earthing contact to the rotating shaft on which the chopper is mounted is not easy. No lubrication can be used and the contact wears. Wear can be minimised by using a smooth shaft of small diameter and by keeping the rotational speed and the brush contact pressure down. The simple approach is to use a pair of thin spring wires resting on the side of the rotating shaft. For low noise a precious metal earthing brush contact is needed. For instruments needed for long continuous operation (more than several months) wear of the brushes presents maintenance problems. In small scale instruments it may not be easy to mount and adjust suitable earthing brushes. At the higher rotational speeds needed for fast response instruments (below say 10ms) a higher contact pressure is needed to avoid contact bounce. This both exacerbates wear of the brushes and so reduces brush lifetime and increases the motor power required.

A 'back to back' fieldmeter approach, that was devised in 1990, overcomes many of the limitations of traditional 'earthed chopper' types of instruments [9,10]. In essence there are two fieldmeters mounted on the same motor drive with the two rotor assemblies electrically connected together but electrically isolated from the motor drive shaft. One fieldmeter observes the external electric field while the other, 'secondary', fieldmeter is in a fully shielded enclosure. By balancing the signal of the primary fieldmeter by an appropriate fraction of the signal observed by the secondary fieldmeter it is possible to fully compensate for the effect of any nett charge held on the dual rotor assembly.

A useful simplification for practical instruments was the realisation that the function of the secondary fieldmeter is, in fact, just to observe the voltage variation of the rotor assemble arising from variations in capacitance of the rotor as it rotates – not the actial level of charge. Figure 3.5 shows the basic arrangement for such a fieldmeter with the rotor assembly, sensing surfaces, motor drive and the phase sensitive detection.

The 'back to back' fieldmeter approach requires near zero end float for the rotor assembly drive (freedom for axial movement) and this requires care in choosing a suitable drive motor. All the normal requirements for good design and construction of fieldmeters (e.g. gold plating surfaces, hiding insulating surfaces, etc) apply with this arrangement.

Immunity to charge on the rotor assembly is easily tested in setting up these instruments by observing the change in output after charge has been added to the rotor, by contact with a battery (e.g. 9V), and then after the rotor has been earthed. The contribution of the signal from the secondary sensing surface is then adjusted to achieve no change of output when charge is added to the rotor.

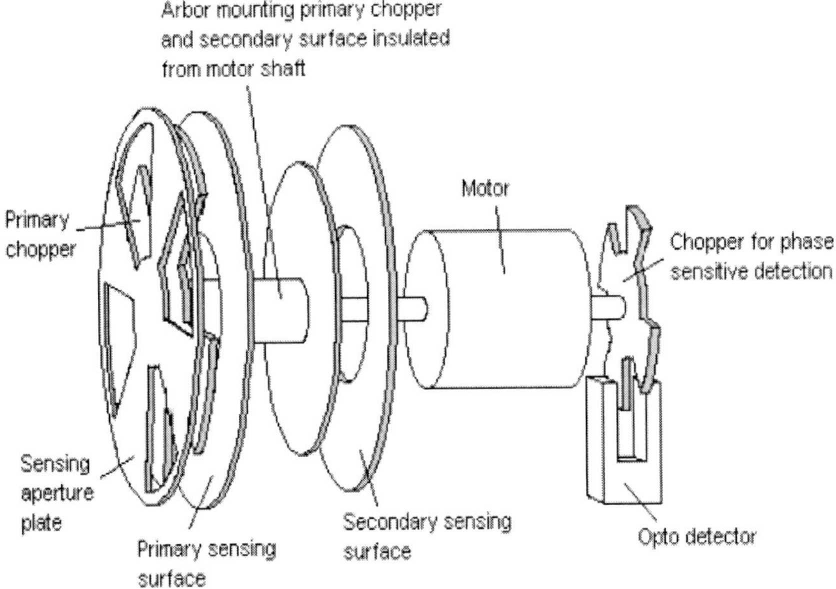

Figure 3.5. 'Field mill' fieldmeter with no earthing of the rotating chopper.

3.2.5. Zeroing Fieldmeters

A sensitive fieldmeter will respond to the presence of an earthed conducting surface of a different metal close to its the sensing aperture. This is a consequence of contact potential difference effects. This is why it is necessary for all surfaces within the sensing region of the fieldmeter to be clean and free of contamination and corrosion. This is best achieved by, for example, gold plating all the relevant surfaces. It is also desirable to avoid close spacing of sensing and chopper surfaces as these exacerbate the electric fields created by any minor differences in surface potentials – for instance due to surfaee contamination.

To check and set the zero of a fieldmeter the sensing aperture needs to be mounted to look into a large clean conducting enclosure connected to the instrument earth providing shielding against any sources of external electric fields. For sensitive fieldmeters (say better than 2kV m^{-1} FSD) it is necessary for the zeroing metal cup to be sufficiently large so that none of its surfaces comes within about 50mm of the sensing aperture. It may even be desirable for the cup surfaces to be gold plated. It is certainly not satisfactory just to lay a metal surface or the palm of the hand over the sensing aperture.

3.2.6. Calibration of Fieldmeters

The design and construction of fieldmeter instruments needs to be such that a definite relationship can be established between the reading, or output, and the electric field to be measured and that it can be understood and used equally by different users. For measurement of electric fields at conducting surfaces this requirement (as specified in BS7506 [11]) is that the fieldmeter has a defined and recognisable 'sensing aperture'. This might be the edge of the casing or a plane surface surrounding the sensing region. For the purpose of setting up and calibrating the electric field sensitivity it is this aperture that is to be considered as the reference surface of the instrument. It is this surface that is to be mounted flush with a large surrounding plane conducting surface that is part of a large two plate parallel plane geometry within which the electric field can be defined with confidence from the spacing between the two plates, d, and the applied voltage, V, as: $E = V / d$.

Formal methods of calibration have been developed and documented [11,12] – and updated versions are outlined in Annex 3. For fieldmeters used to measure electric fields in the atmosphere or in large volumes no mounting surface need be involved and alternative procedures for calibration are appropriate - as discussed in Chapter 4.5.

3.2.7 Factors Determining Performance of Field Mill Fieldmeters

3.2.7.1. Introduction

Making a fieldmeter that responds to the strength and polarity of an electric field is not too difficult and many people have done this in various configurations. What is more difficult is to make fieldmeters with high sensitivity, good zero stability, low noise, fast response, long operational life and immunity to local environmental conditions – and to provide several of these features together in a single instrument. The following paragraphs outline some of the approaches that have been developed to tackle these objectives.

3.2.7.2. Earthing Rotating Choppers

Rotating choppers are usually earthed by brushes in rubbing contact with the rotating shaft. Neither sleeve nor ball bearings provide good earthing – basically because the lubrication involved isolates the shaft and bearing surfaces. For low electrical noise to the fieldmeter input it is desirable that the contact brushes are made of a precious metal alloy spring wire (fairly hard) and resting with just a few grams force (say 5-10g) on a small diameter section of smooth shaft. It is best to use a couple of brushes. The brushes wear with operation and need to be replaced from time to time. Experience has been that several months of continuous operation can be expected from a pair of brushes. If high rotational speed is needed to provide a fast response output (better than about 10ms) then the brush pressure will need to be increased to avoid the brushes bouncing during rotation and thereby creating noise through the consequent isolation of the chopper assembly. This higher pressure will decrease the lifetime of brush operation and increase the drive power required.

3.2.7.3. Zero Stability

Contact potential differences between chopper and sensing surfaces will give rise to a finite output signal even when the external electric field is zero. Such effects are best minimised by using stainless steel or gold plating for all surfaces around the sensing region so there are no changes in contact potentials likely to result from corrosion in long term studies. It is also advantageous to have sizeable gaps between chopper and sensing surfaces to minimise the electric fields created by whatever residual electrochemical potentials are present. However, if the gaps are large compared to the size of the chopper sectors then there will be a sizeable reduction in depth of modulation and so a reduction in fieldmeter sensitivity.

A major factor affecting the zero in practical measurements is the influence of any contamination on chopper or sensing surfaces. This is a particular problem when measurements are made in situations where airborne particles or highly insulating powder particles are around. If these are deposited on surfaces in the sensing region then appreciable zero offsets can arise. The best way to combat this is to provide a flow of clean, particle free, airflow through the sensing region and out of the sensing aperture to minimise ingress of contaminants. It is also desirable to check the zero reading from time to time – see Section 3.2.5 above.

Zero stability can also be affected if signals can be picked up (via capacitance or inductive coupling) from the motor or the motor leads. While it may be feasible to balance out such signals when the instrument is initially set up they may change with time as motor commutation and bearings change with wear in long term use. For fieldmeters without earthing of the rotating chopper (Section 3.2.4 above) this also requires the capacitance between the chopper assembly and the motor shaft to be kept very low. It also requires that there is negligible coupling between the motor shaft and the secondary sensing surface as this would be a signal that could not be nulled out by compensation against rotor charge. It is desirable to have magnetic shielding between the motor and the signal connections and processing circuits to minimize inductive signal pick up.

3.2.7.4. Precision and Sensitivity

Achievement of a stable and precise relationship between signal output and the electric field at the sensing aperture requires good mechanical stability of all surfaces in the sensing region. There should be no exposed insulating surfaces in the sensing region – so insulation mounting the sensing surfaces needs to be secluded. There needs to be very little end float (freedom for axial movement) in the chopper drive system otherwise there will be a change in the depth of modulation as the chopper moves relative to the sensing surfaces – for example due to a change of attitude in practical use. Where the chopper assembly is mounted directly on the motor shaft this means choosing motors with preloaded ball race bearings that eliminate end float. End float needs to be especially small with fieldmeters without earthing of the rotating chopper (Section 3.2.4 above).

Good precision requires careful attention to a variety of factors affecting signal to noise ratios throughput signal processing. This is of course a particular area of attention at the front end – and attention is needed to sources such as motor noise (3.2.8.3 above).

There are a number of ways the induced charge signals from the sensing surfaces may be processed, from the voltage developed across an input capacitor or by a direct virtual earth charge sensing circuit – as discussed in Section 3.2.2 above. Measurement by the current flow is generally not a good approach as the current will vary with the speed of rotation and this may vary in battery powered instruments or with bearing or commutator wear. Voltage measurement has the disadvantage that it requires the input leakage resistance of the sensing surface to be very high – particularly so for high sensitivity. This makes operation sensitive to contamination and moisture on the insulation mounting the sensing surfaces and the input to the preamplifier circuit. Use of the virtual earth charge measurement circuit minimises susceptibility to leakage from the sensing surfaces because these remain at earth potential. Leakage across the feedback capacitor is more likely to remain stable as it will be within the same protection as for the signal processing circuits. It may be useful to use guard rings or to make electrical connections via air linkages to avoid leakage on, or in, the surface of circuit boards.

The best way to process the alternating waveform signals after amplification is phase sensitive detection. This involves switching the signals accurately in phase with alignment of the chopper and sensing surfaces, or sensing aperture, and is followed by signal filtering to attenuate signals at and above the chopping frequency. Salen and Keys filtering is an appropriate approach. The output then gives low noise and a linear response.

The signal for operation of the phase sensitive detector may be obtained in various ways - for example, an infra-red opto detector or a magnetic reluctance sensor operating in conjunction with a secondary chopper and arranged to be in exact phase relation to the observed electric field signals. After detection the half-wave signal should be exactly balanced to be 180 degrees. It needs to be noted that infra-red detectors may be susceptible to tungsten lamp and sunlight illumination – a problem that is avoided with magnetic sensing. An alternative approach can be to use an electronically commutated motor for chopper rotation and derive the phase sensitive detection drive signal from a rotation logic signal.

The signal from the sensing surfaces is the average of the signals received from each sector of the sensing surfaces. If the detector for the phase sensitive detection signal is derived from a localized sensor there is opportunity for any irregularities in this localized signal to create variations in phase detection timing over the cycle of chopper rotation. This can create noise in the output signal at frequencies up from the timescale of rotation of the chopper and this will limit the frequency response for low electric field values. A good solution to this is to use a phase locked loop circuit to smooth out the timing of the phase switching over a time suitably longer than the time for chopper rotation.

While it is true that the signal from the sensing surfaces is the average of the signals received from each sector it is best that the chopper discs rotate concentrically and in a plane with no swashplating. It is also useful for the chopper disc to have a rim to minimize the generation of signals by variations of capacitance coupling between chopper sectors and irregularities in the surfaces surrounding the chopper. A rim also makes it easier to manufacture, handle and to use chopper discs of thin material.

3.2.7.5. Response Time

The response time of a 'field mill' type fieldmeter is determined primarily by the frequency of modulation of the electric field at the sensing surface. With rotating chopper fieldmeters the frequency of modulation relates to the number of chopper sectors and the speed of chopper rotation. Rotational speeds of 10,000 to 15,000 rpm (170 to 250 rps) are reasonable to achieve over extended periods of operation in practice. The number of sectors is limited by the need to retain a sensible ratio between the aperture of chopping sectors and the spacing to the sensing surface, or the sensing aperture, so there is adequate coupling and modulation of the external electric field to the sensing surface. With say 10 sectors this means that the chopping frequency is not likely to be greater than 1.7 to 2.5kHz. On this basis one might expect to be able to achieve a response frequency up to say 0.5kHz.

When the response time of the signal output only needs to be slow compared to the time for chopper rotation then it is easy to filter out signal variations arising from irregularities in chopper geometries, etc. affecting phase sensitive detection. However, where the timescale of response needs to be comparable to, or shorter than, the time for rotation of the chopper and where a high sensitivity is also required then special care is needed. Attention is needed to each of the points made in section 3.2.7.4 above. To get the maximum frequency response from the fieldmeter the output needs to be filtered with as high a roll-off frequency as is compatible with achieving adequately low noise from the chopping frequency in the final output. Two stages of Salen and Keys filtering in series may be desirable.

3.2.7.6. Adverse Environmental Conditions

Maintenance of performance in wet and dirty operating conditions is not easy and requires careful design and construction. To avoid water bridging between sensing surface sectors and nearby surfaces it is wise to have a separation between the sensing surfaces and nearby surfaces greater than 5mm. This means that fieldmeters for situations where water is likely to get into the sensing region need to be fairly large. It is also necessary that insulation mounting sensing surfaces provide long tracking paths and that signal processing circuits are enclosed and suitably protected. The influence of electrical leakage across insulators mounting the sensing surfaces is minimized by using virtual earth charge measurement of the signals – as shown in the circuit in Figure 3.3.

The surfaces of chopper and sensing surfaces should all be the same metal and not be susceptible to corrosion to avoid the influence of electrochemical potential differences between the surfaces. Gold plating is a good way to achieve non-corroding surfaces. Stainless steel should also be suitable – but preferably all the surfaces should be the same. While stainless steel can be gold plated this may not remain well adhered during long term operation is adverse environmental conditions.

Despite best precautions situations can arise when performance is degraded in adverse operating conditions for example by the build up of debris or the entry of small animals or insects. To give confidence against such risks in long term studies it is desirable to provide some form of operational health facility. This may conveniently be in the form of a defined independent local source of alternating electric field whose amplitude can be monitored through the fieldmeter system [13,14,15]. Two main approaches have been developed for monitoring the operational health of fieldmeters in adverse environmental conditions – for instance in the measurement of atmospheric electric fields. The first approach involved applying a defined amplitude low frequency modulation (around 1Hz) of the potential of the whole fieldmeter structure [14]. The second involved application of a defined amplitude alternating voltage to a shield electrode mounted near the sensing aperture [15]. The frequency of the alternating voltage in this case was above the response frequency of the fieldmeter output and was phase locked to the chopping frequency. The observed electric field signal was phase sensitive detected at this frequency. In both cases the amplitude of the operational health signal observed was compared to the expected value so that an output could be generated and recorded showing the operational health status of the fieldmeter.

3.3. Measurement of Charge

3.3.1. Introduction

The nett charge on moveable items, conducting or insulating, that are separated or isolated from ground is best measured using a Faraday Pail. This is relevant to measurements on solids, to pieces of material, to falling particles and to liquids and powders exiting from a pipe.

The measurement of the charge transferred in static discharges is a separate issue and is discussed in Section 3.7 below.

3.3.2. Faraday Pail

The basic concept of a Faraday Pail is a conducting chamber in which the electric field created by charge on the item to be measured is all captured on surfaces of the pail - and none couples to the world outside. In this situation if the chamber is isolated from earth then all the charge introduced into the pail also appears as an exactly equivalent charge on the outside of the pail, where it can be measured. It is not necessary for the charge introduced to actually flow to the inside of the pail - so the method is equally useful for charge on insulators as on conductors. It is however necessary to note that the charge on an item or material put into the pail must be allowed to couple fully to the pail. Hence, the charged item or material must not be, for example, mounted on or close coupled to an earthed retaining support that would affect free coupling to the pail. Nor may it be mounted on insulation if this holds charge, as this charge could also couple into the pail and affect observations.

The Faraday Pail may be used equally well to measure charge on individual items, on a collection of items, on a volume of liquid or on particles suspended or moving in air or a gas.

Basic features of a Faraday Pail are illustrated in Figure 3.6. The Faraday Pail itself needs to be provided with an earthed conducting shield to prevent charges being induced directly on the pail by electric fields from charged surfaces in the nearby surroundings.

The pail needs to be designed and built with the following two main features in mind [11]:

- the geometry of the pail shall ensure that all the charge introduced into the pail couples to the pail and none to the world outside.

  This means that a closed ended pail needs to be fairly deep (a height to diameter ratio of at least 1.4 to 1 with the charge retained within the lower 40% of the height). Similar considerations are needed to ensure full coupling of contained charge within double open ended and other shaped vessels
- the pail needs to be well shielded against the influence of charges in the nearby vicinity.

The pail may take many forms so long as it satisfies the above basic requirements. For example to measure the charge on individual raindrops the 'pail' might be a vertical cylinder open at both ends through which individual rain drops could fall.

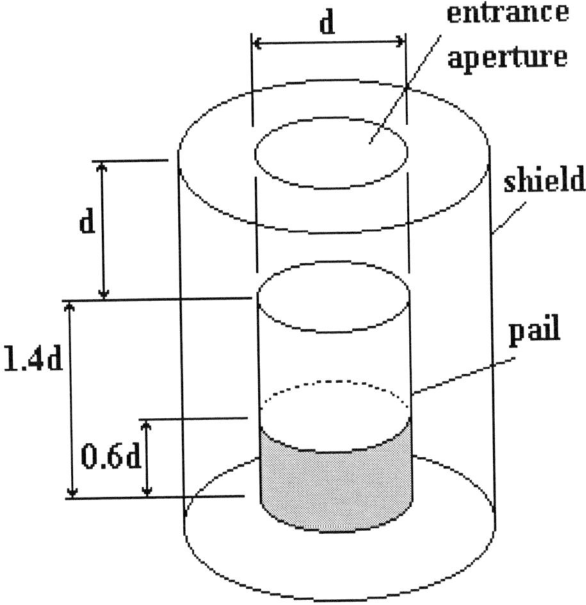

Figure 3.6. Basic features of a Faraday Pail.

3.3.3. Charge Measurement

The charge induced on the pail is most appropriately measured using the virtual earth charge measurement circuit - shown in Figure 3.3. In this circuit capacitance feedback around the operational amplifier keeps the input at earth potential. In this way all the charge on the source appears on the feedback capacitor and the quantity of charge is the product of the feedback capacitor value and the output voltage.

Measurement of charge by the voltage developed across a capacitor is sometimes used. This it is not preferred because the increase in voltage on the capacitor and pail will introduce an error in charge measurement. If charge is measured by the voltage developed across a capacitor of known value then there is charge sharing and the voltage developed on the measurement capacitor depends upon the capacitance of the charge source. If C_s is the capacitance of the source of the charge and C_m the capacitance of the measurement capacitor then the voltage developed, V_m, will be:

$$V_m = Q_s /(C_s + C_m)$$

So long as $C_m >> C_s$ the voltage developed on the measuring capacitor will be a reasonable measure of the quantity of charge entered in to the pail. The input leakage resistance needs to be sufficiently high that the RC time constant with the measurement capacitor is long compared to the timescale of measurement or recording of observations. An electrostatic voltmeter may be useful here for voltage measurements. Zeroing for the voltage measurement approach involves simply connecting the pail temporarily to earth.

The virtual earth charge measurement approach is the most appropriate approach. The basic limitations are the maximum voltage swing available and the maximum rate of charge flow into the pail. If the rate of charge inflow exceeds the output current capability of the amplifier stage then errors will arise in charge measurement. The virtual earth circuit aims to maintain the input at earth potential, but if moisture and contamination are present on the input insulation then battery voltages generated will cause zero and reading values to drift.

Zeroing of a virtual earth measurement circuit is achieved by shorting the feedback capacitor – <u>not</u> by earthing the input connection! If it is desired to make measurements with stability and resolution down towards the picocoulomb range it is best to make circuit connections in air and off the surface of a printed circuit board - or to use guard electrodes around sensitive connections. Isopropanol make prove a useful cleaner for circuit components.

The performance of a Faraday Pail system for charge measurement may be checked and calibrated by charging a known value capacitor to a known voltage so as to inject a suitable and defined quantity of charge. It is important that a good quality capacitor is used and that the charging voltage is at least several volts – to minimise risk of influence by electrochemical effects. The charged capacitor approach is not so appropriate for calibration into the picocoulomb range because the value of a suitably small capacitor is likely to be affected by proximity effects. An alternative approach is to create a defined flow of current and to switch this to the charge measurement input for a defined period of time. This approach is most appropriate for use with virtual earth charge measurement circuits. A defined and stable flow of current can be achieved from a referenced voltage source and a defined precision resistor to earth. The flow of this current at the earth connection point may be electronically switched to flow into the input of the virtual earth measurement circuit rather than to earth for a defined period of time. As the source current flow is continuous there is no influence from any distributed capacitance in the precision resistor or its connections. The electronic switch and the layout of the circuit need to be chosen to give negligible charge injection at operation. The period for current flow can be derived with high accuracy and stability by scaling down appropriately from a quartz crystal oscillator. Each of the 3 factors determining the quantity of charge output (voltage, resistance and time) can be formally calibrated with reference to National Standards. Formal calibration procedures are described in Annex 3.

3.4. Assessment of Materials

3.4.1. Introduction

Static electricity arises on the surface of materials by contact or rubbing actions with other material surfaces or by the fracture of surfaces. The risks and problems created by static electricity and the opportunity for its constructive depends on four main features. The relative importance of each depends on the area of application.

1) voltages arising on surfaces after these are contacted or rubbed by other surfaces
2) ability of surfaces to drain charge away from conductors in contact
3) ability of materials to provide shielding against electric field transients
4) ability of a material to support an incendive electrostatic discharge

The first of these, the surface voltage that arises from charge retained on a material itself, is the root cause of most electrostatic problems and opportunities for applications. It is the surface voltage created that is responsible for the electric fields at nearby items and is responsible for attraction of dust and debris, for the clinging of thin films, for shocks, for the occurrence of incendive discharges and for risks of damage to microelectronic components and systems.

The ability of materials to drain charge from conductors in contact and away to earth is appropriately covered by resistance measurement [1,4] - for example for footwear and flooring to control body voltage during activities such as walking. Shielding and the support of incendive discharges are separate topics that require their own appropriate methods of measurement. The measurement of shielding performance is discussed in Section 3.5 and assessment of the incendivity of electrostatic discharges in Section 3.6.

The voltages that arise on surfaces after these have been contacted or rubbed together will be influenced by the materials involved via the triboelectric series. In practice, where materials and surface conditions cannot be well defined it is not likely that the triboelectric series will provide much useful guidance. The voltages that arise are determined by two main factors: first, how quickly charge can migrate over the surface compared to the time it takes for the surfaces to separate after contact; and second, the capacitance experienced by the charge on the surface. The measurement of these parameters is the subject of the following discussion.

3.4.2. Philosophy of Assessment

Traditionally the suitability of materials to avoid problems from retained charge has been assessed by measurement of resistivity. The view expressed in present Standards is that so long as the resistivity is below 10^{10} ohms there will be no problems [1,4]. This not an appropriate philosophy and does not work for many modern materials.

Composite and inhomogeneous materials may include both relatively insulating and relatively conducting parts. A simple example is the fabric used in cleanroom garments. The basic fabric is usually polyester, which easily acquires a high surface voltage when rubbed. In practical use the fabric usually includes a pattern of conductive threads. If these are surface conductive threads then low resistivity values will be observed when the electrodes for surface resistivity measurement contact the conductive threads. It is clear however that such measurements provide no information about the surface voltages that will arise on the relatively insulating parts between the threads when the fabric is rubbed. It is on these areas, of course, where electrostatic charge will be retained. Resistivity measurements will hence provide no guidance on the effective surface voltages that will arise when such fabrics are rubbed in practical use [16].

The logical way to assess materials for the risk presented by retained surface charge is to put some charge on the material, by a rubbing or scuffing action, measure what surface voltage is created per unit of charge, and to measure how quickly the surface voltage that was created falls as the charge moves away over the surface. These two measurements can then be compared to appropriate threshold acceptance criteria. Standard numerical values may be appropriate for general application but specific values may be more appropriate for particular areas of application.

It is appropriate to relate the surface voltages to the quantity of charge transferred. This avoids dependence on such variable factors such as the speed and pressures of rubbing that may easily vary with the method of rubbing.

Studies on a variety of materials by the 'scuff charging' approach have been carried out and published [17,18,19,20]. Such studies have indicated that decay times down below around 0.25 second are needed to limit surface voltages to low values against manual rubbing actions. This is in line with the expected need for the decay time to be small compared to the time for manual separation of contacting surfaces. These studies have also shown that maximum initial peak voltages can be held to low values if the charge experiences a high 'capacitance' at the surface [17,18,19,20]. The 'capacitance' experienced by charge on a surface is a feature of the dielectric constant of the material and, in particular, the proximity of any conductive features within or immediately under the material. It does not seem practical to directly measure the capacitance experienced by surface charge. The problem is that in practical charging the charge density varies over the area charged so there is no defined boundary for the charge. The term 'capacitance loading' has been devised to describe the capacitance effect experienced by surface charge. 'Capacitance loading' is defined [17,20] as the ratio of the effective capacitance observed with the sample surface divided by that observed with a thin layer of a good dielectric film – for instance cling film. This is expressed by the surface voltage created per unit of charge deposited on a thin layer of a good dielectric surface divided by the surface voltage created per unit of charge with a similar spatial distribution on the test material.

Tribocharging is not an easy way to make measurements [17,21]. The charging action needs to be completed within quite a short time and the item used to achieve charging needs to move fully away very quickly so it does not affect measurement of the surface voltage created. An arrangement that has been used in studies on fabrics is illustrated in Figure 3.7 below.

The quantities of charge easily and quickly transferred to a surface in experimental studies may be quite a bit less than could arise in practice with higher speeds and pressures of contact. Thus with materials that provide a high capacitance to surface charge it is necessary to be able to measure quite low surface voltages and to do this with a good time response. It is also necessary for the charging action to avoid mechanically impacting the fieldmeter used for surface voltage measurement instrument and for the charged rubbing material to be quickly removed well away to prevent any influence to the fieldmeter [17].

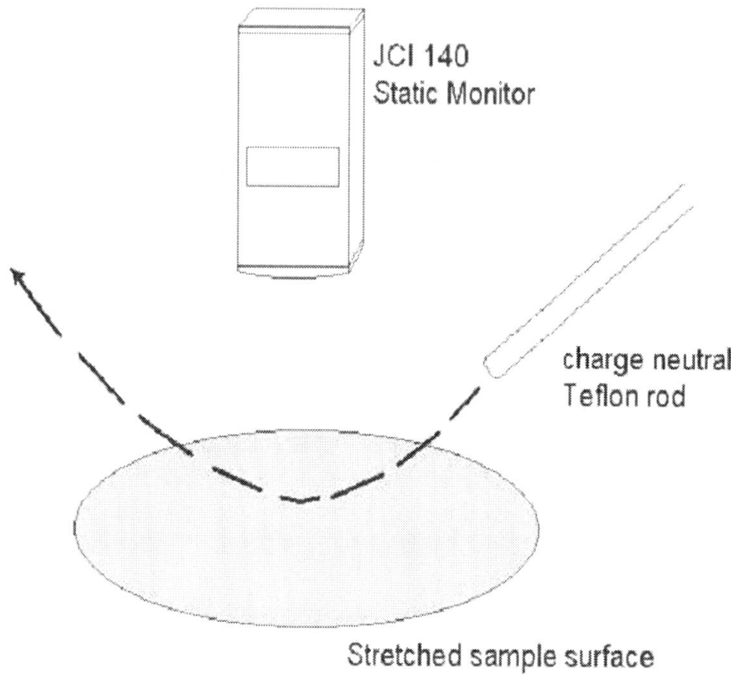

Figure 3.7. Experimental arrangement for studying scuff charging with fabrics.

The alternative to tribocharging is to use corona charging to place a local patch of charge on to the surface of the material [22]. Studies with a variety of types of material have shown that measurements of decay times and of capacitance loading with corona charging give a reasonable match to these characteristics measured with tribocharging [19,23]. Corona charging provides the basis for compact and easy to use instrumentation that can be used directly on a wide range of materials - including powders and liquids.

3.4.3. Comments on Methods for Charge Decay Measurement

Several methods for measurement of 'charge decay' are described in various 'Standards' and in published literature [11, 22-37] and comments on these have been published [38]. The methods are very different in principle, do not in general agree with each other and in general do not provide the information likely to be useful for uninitiated or unskilled users because they are susceptible to the construction of the materials tested.

It has been proposed [38] that methods of charge decay measurement need to satisfy the following requirements:

- give results comparable to measurements with practical tribocharging actions for a wide variety of materials
- that the material is initially at near earth potential and is subject to a charging action in a central region near a surface voltage sensor.
- measurements are made of the surface voltage created by the charging action and how this decays with time
- surface voltage measurements are made directly of the area charged and without contact
- measurement should be made on the same side of the material as that charged
- be independent of constructions or features of materials
- make zero or minimal modification of the material and decay characteristics by the conduct of the test (and this should also apply to tribocharging methods).
- the method should be easy to use and to interpret by non-specialist staff
- suitable equipment should be easy to construct and/or be commercially available.
- approaches should also be backed by peer reviewed published papers describing the equipment and giving supporting experimental measurements by different workers. Where appropriate these papers should be referenced in the 'standard' document.
- the measuring instruments should be able to be formally calibrated

The following notes describe a number of methods of charge decay measurement in use with comments on their strengths and limitations. It may be argued that many of the methods mentioned are 'well established', in common use and are specified in 'Standards'. However, this does not make them right! It is hoped the following comments will clarify their strengths and weaknesses.

Corona Charge Decay

The corona charging approach to charge decay measurement offers many practical advantages and has been implemented in compact easy to use commercial instrumentation [22]. It is in use by non-specialist staff with many types of materials in a wide variety of industries around the world. Corona charge decay is included in a couple of formal standards documents [11,26]. A new test method Standard has been proposed [38,39] as shown in Annex 2.

This test method can be used in conjunction with measurement of the quantity of corona charge transferred to the test surface to give values for the 'capacitance loading' effect experienced by charge on the surface tested [17,20,23,24]. This is of practical relevance as it shows the surface voltages expected from practical quantities of tribocharge [20,23]. Corona charging gives a generalized way to assess the suitability of materials, including films, layers, liquids, powders, small items and installed surfaces.

Studies have been reported that show comparability between corona and tribocharge decay times for a variety of materials [19,23]. These have shown that it is important with short decay time materials to compare decay rates at comparable times after the end of the charging action [19]. Lack of any significant modification of the surface by the action of corona ionization has been shown [25] by constancy of charge decay performance at a single location from an initial low corona charge exposure through a high exposure and back to low exposure.

The method needs to be checked more extensively by different workers – in particular for variation of characteristics with the area of surface charged and with further comparisons to tribocharging results.

Federal Test Standard 101C

Federal Test Standard 101C Method 4046 [27] has been around for many years and has been subject to a number of comments and refinements [29,30]. The basic approach involves mounting a 5" long and 3" wide strip of material between supporting clamps in front of a fieldmeter. A voltage of 5000V is applied to the clamps and the build up of fieldmeter signal is observed to achieve a reading equivalent to the applied voltage. The clamps are then earthed and the decay of the fieldmeter signal observed and timed. It is noted in the specification that the method should only be used for 'homogeneous materials' - but no guidance is given on how to recognize such materials!

Comparative tests have shown much shorter charge decay times by FTS 101C than are observed with corona charge decay measurements with many practical materials [28]. These tests did confirm that comparable results are obtained with truly homogeneous materials. It was concluded that FTS 101C basically responded to the fastest route for charge movement in the layer of material, whereas corona charge decay showed the slowest way that charge moved on the surface of materials. The method of charging in FTS101C does not ensure full charging of the more insulating components in material within a relatively conductive matrix or grid whereas corona charging, like tribocharging, provides charging whatever the nature of the surface. The FTS101C method requires use of a cut sample area and cannot be used on practical surfaces. Equipment is available commercially.

ITV Denkendorf

ITV Denkendorf developed a tribocharging method (ITV-TEV) in which a nearly vertical strip of material is held under tension between two earthed clamps. The strip is rubbed between a pair of offset polythene rollers as these are moved down the strip under tension [30]. The rubbed area is held stable in front of a fieldmeter to observe the initial peak voltage and the rate of charge decay. The principle of the method seems sound. It is however only applicable to flexible layer materials and to materials with decay times longer than several tenths of a second due to the mechanical time needed to moved the charging rollers. Rather different behaviour is observed with fabrics in the warp and in the weft directions. The equipment is not now available commercially.

NASA

A tribocharging method for testing layer materials has been developed at NASA by Gompf [21]. This uses a rotating Teflon brush to tribocharge the sample surface that is earthed around its edge. At cessation of charging the mounted sample is quickly swung in front of a fieldmeter so the initial peak surface voltage and the rate of decay of this voltage can be measured. This seems a good, valid and useful approach. Results are reported to correlate well with safety experience at NASA. There also seems reasonable correlation with two other test methods [31]. The approach had been limited by using an induction probe type fieldmeter, rather than a field mill (used more recently), and by the time taken to move the sample to the position of observation at the cessation of rubbing. Use of an induction probe limits the sensitivity for low surface voltages and the length of decay times that can usefully be measured. It is also, of course, limited to layer type materials that can be presented as cut samples and those not likely to be damaged by the fairly vigorous tribocharged rubbing action used. There is also a limitation in the minimum decay times that can be measured by the mechanical delay between the end of charging and the start of surface voltage observations. This equipment is not available commercially.

A modification has been proposed to the above approach [32] to try to simulate the risk from the charging of an unearthed person's body while wearing personnel protective clothing. An isolated pick up disc has been mounted as a backing support for the sample with 220pF capacitance to earth. The idea is that the electrostatic energy picked up by this disc will represent the energy available to create risks of ignition. However, if the sample is mounted on an earthy support then the presence of conductive threads in the test fabric could diminish the quantities of charge observed because of shielding effects. It needs to be recognized that risks may also arise if there are local areas of high voltage on garment fabric surfaces relative to the body – although the incendivity of such discharges will be affected by other characteristics of the fabric.

BTTG

A corona charging method is used at British Textile Technology Group (BTTG) for testing fabrics (Shirley Method 20). A 300mm diameter disc of material is held under radial tension on a circular conducting frame isolated from earth connection. The material is charged by a cluster of corona discharge points near the centre of the disc on one side and a fieldmeter observes the surface voltage on the opposite side. After the surface has been charged the mounting frame is connected to earth.

The observed charge decay behaviour after the mounting frame is earthed needs to be assessed with caution. First, observations with the fieldmeter are made on the opposite side to that charged. For materials including relatively conducting components within their structure (for instance conductive threads) there will be a shielding effect between the charge and the observations so observations may not directly relate to the behaviour of surface charged. Second, as for FTS 101C, the initial reduction of observed signal will be strongly influenced by capacitive coupling via high conductivity components in the material. It is then a matter of judgement as to when observations relate to the decay of surface charge on the fabric itself. It is only applicable for layer type materials that can be presented as cut samples. This equipment is not available commercially.

BTTG has also developed a corona charging method for testing whole garments (Shirley Method 137 for Charge Decay Time Measurement on a Full Garment [33]). This involves using a corona discharge to charge an area of the garment, while the garment is hung up vertically from insulated supports. The variation of the potential at the charged area is observed from the time the corona charging electrode system is removed with four different test procedures. Charge decay behaviour is observed with four test procedures: charging with the garment unearthed and then earthed via the cuff and then the ankle area and charged while continuously earthed via the cuff and then via the ankle area. This equipment is not available commercially. Only the later two test approaches seem appropriate.

'Scuff' Tribocharging

A simple method for studying tribocharging has been developed [17,23,24]. It involves scuffing the middle of a stretched area of film or fabric with a charge neutral Teflon rod. A fieldmeter above the struck area shows the initial peak voltage created by the charging action and the rate of decay of this charge. The approach is illustrated in Figure 3.7 above. Measurement of the quantity of charge transferred at the scuffing action is made by putting the struck end of the initially charge neutral Teflon rod into a Faraday Pail. In more recent studies localized scuff charging was provided by a ball dropped on to an inclinded test surface with the ball bounced directly into a Faraday Pail for charge measurement [34]. Scuff charging has been used in studies with inhabited cleanroom garments where the charge transferred was measured either by the charge retained on an initially charge neutral Teflon rod [15,20] or by having the test subject stand on a plate connected to a charge measurement circuit [34]. Values for the quantity of charge per unit of initial peak surface voltage also show the capacitance effect experienced by the charge on the surface.

The main value of the method is probably to provide a tribocharging reference against which other test methods may be compared. It only really seems suitable for laboratory research type studies with film and planar surface materials. It does not seem appropriate for commercial instrumentation. No studies have yet been reported on results with charging of large areas or with repeated charging.

STFI

A method for assessing materials has been developed by STFI [35]. A sample layer of material is clamped over the end of an earthed ring electrode. A step function high voltage pulse is applied to a disc electrode near the underside of the sample and the form of the signal observed by a pick up disc electrode near the far side of the layer is recorded. With careful interpretation these observations seem to relate to the risk of occurrence of incendive electrostatic discharges from charged surfaces of the sort of fabrics used in Type C FIBC (Fabric Intermediate Bulk Containers). These fabrics include a pattern of conductive threads that are required to be earth bonded in their application.

The method is not suitable for measuring the surface charge decay capabilities of layer materials. For instance it will evidently not measure the decay of charge for a layer of plastic coating on a metal sheet!

The method is not appropriate for measuring the surface charge decay capabilities of materials. The material is 'charged' by induction so components with long decay times will only be charged to a low level. This means that only small surface voltages will be available for surface charge decay time measurements. A field mill type fieldmeter is needed, rather than an induction type, to monitor charging and charge decay. If conductive threads or internal high conductivity components are included in the material tested their influence will depend on their resistive and capacitance coupling to the earthed mounting. This will affect observations. The method cannot be used with materials against an earthed backing.

The response time of observations to the fast rising applied electric field gives indication of the effective conductivity within the material providing shielding performance. It seems very reasonable that this has a relation to the opportunity for drawing incendive electrostatic discharges from the material surface [45]. Resistive and capacitance coupling at the earthed mounting of the sample will affect observations. It has been indicated that this equipment is to be commercially available.

Charge Plate Monitor

Observations are often made using a metal electrode connected to an electrostatic voltmeter in contact with the middle of an area of test material [36,37]. A popular approach is to use a 'charge plate monitor' (an instrument designed for assessing the ability of air ionization to remove charge from surfaces).

This approach may be useful for assessing how quickly charge may be removed from a conducting item in contact with a material - such as a person standing on flooring. It does NOT, however, measure the ability of a material to self-dissipate charge on its own surface. As with FTS 101C etc, the reason is that observations are strongly influenced by any conduction linkage to the fastest routes for charge migration and lack of fair representation of the influence of slow charge migration routes of relatively insulating components. There may also be limitations on long decay times by the electrical leakage of the insulation mounting the plate. There is also an uncertain influence from the capacitance loading of the contacting electrode affecting RC time constant evaluation. Equipment is available commercially from several sources.

Comment

Eight methods have been described above for measurement of 'charge decay'.

- Four of the methods are not appropriate for assessing how quickly charge can dissipate on a material itself – FTS 101C, BTTG, STFI and the Charge Plate Monitor. The basic problem with these methods is that their observations are dominated by the fastest route for charge migration in the material, and they do not show the behaviour of charge retained on the surface of the material - as after tribocharging. No attempts have been reported to relate such methods to tribocharging behaviour.
- None of the methods that start with the material already charged and tested by earthing the test area boundary can be considered valid because this situation does not match practical application experience.
- Of the tribocharging methods only the NASA, ITV Denkendorf and the 'scuff charging' approaches start with the sample initially earthed and with an earthed boundary. None of the three methods described based on tribocharging are available commercially.
- Corona charging provides the basis for easy to use instrumentation and gives results with a variety of materials that match to experience with tribocharging. It provides information both on how quickly charge can self-dissipate on a material and on the surface voltage likely to be generated by various quantities of charge. Instrumentation is commercially available.

The conclusion drawn from the above comments is that while assessment of performance with tribocharging must remain the reference benchmak for the performance characteristaics of materials corona charging provides a valid and practical alternative. The corona charging approach provides the basis for instrumentation with greater versatility, suitability for a wide variety of materials, ease of use and of commercial availability.

3.4.4. Corona Charge Decay Measurement

Corona charge decay measurements are made using a high voltage corona discharge to deposit a small patch of charge on the material to be tested with a fast response 'field mill' electrostatic fieldmeter used to measure the surface voltage created by this patch of charge and how quickly this voltage falls as the deposited charge migrates away.

A physical arrangement for corona charge decay measurement is shown in Figure 3.8. This arrangement has been described in a number of papers [22,23,24,27,28] and is included in a British Standard [11] and an international standard [26].

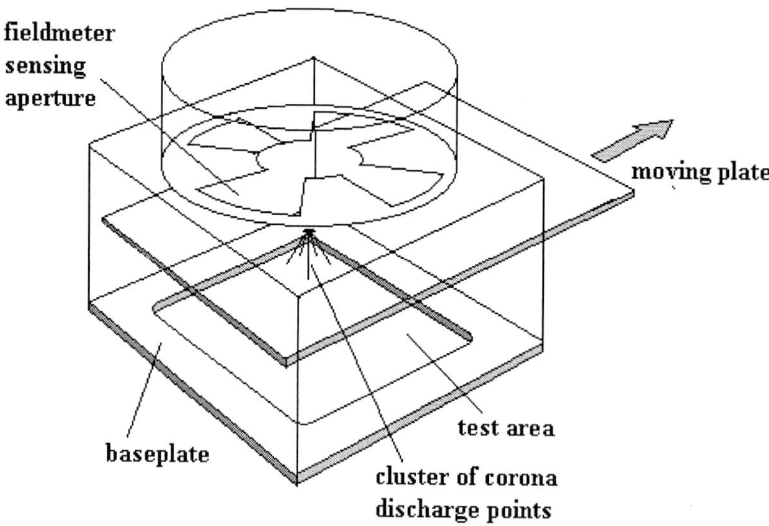

Figure 3.8. Arrangement for measurement of corona charge decay.

The corona discharge is created by a brief pulse (typically 20ms) of high voltage (in the range ±2.5 to ±10kV) applied to a cluster of discharge points mounted on a moveable plate a short distance above the surface to be tested. The corona discharge deposits a local patch of charge on to the surface without contact. As soon as charge has been deposited the moveable plate is moved quickly away (in 20-30ms). A fast response 'field mill' electrostatic fieldmeter, that has been shielded by the earthed upper surface of the plate, is used to measure, without contact, the voltage developed on the surface by this deposited charge and how quickly the voltage falls as the charge migrates away. If measurement is made of the quantity of corona charge deposited then

the effective capacitance experienced by the charge on the test surface can be derived from the initial peak surface voltage created by this quantity of charge – as is discussed later. A full description of the design and test procedure for corona charge decay measurement is given in Annex 2 [40]. A paper has been published on the relevant background [39].

Important practical features of charge decay measuring instrumentation are:

- that the times over which the corona charge is deposited and the time taken by the plate carrying the corona discharge electrodes to move fully out of the way for full view of the test surface by the fieldmeter shall both be short compared to the minimum decay time to be measured. It is difficult to move the plate carrying the corona discharge points fully out of the way of the fieldmeter observations in less than 0.02s. In this situation the minimum decay times that can sensibly be measured will be around 0.05s.
- To avoid problems from static electricity in practical situations it is desirable that charge decay times are short compared to the times for separation of surfaces after mechanical tribocharging actions. This means decay times should generally be less than ¼ second. To cover this, and faster manual actions, it is desirable to be able to measure decay times (see below) to 0.05s less.
- Experience from studies with a wide variety of materials shows that the form of charge decay curves a) rarely follow an exponential form, and b) that the form depends little, if at all, upon the quantity of charge transferred to the surface and hence the level of initial peak voltage generated. In this respect it is not important what initial peak voltage is achieved by the corona charge deposited – only that the initial surface voltage is sufficient for good quality measurements.
- In tribocharging situations it takes a finite time for the contacting surfaces to separate and for the separated charges to create an influence on items nearby. In manual actions this time is typically 0.1s or so. In the light of this it is appropriate to assess materials on the basis of a decay time to a set % of the surface voltage that is provided at say 0.1s after the end of corona charge deposition. Decay time measurement described in Annex 2.
- The initial peak voltage achieved may vary greatly between materials. Because the form of the decay curve is not an exponential it is NOT appropriate to measure decay times from some predefined voltage

level to a set % of this or to some other predefined end point voltage level.

- The % value chosen for the end point of timing may conveniently be 1/e (37%) and/or 10% of the initial voltage. There is no absolute significance to be attached to the 1/e and 10% levels. These just provide convenient numbers which enable the performance of materials to be compared. The fraction chosen should be clear in the reporting of results. The advantage of using a 10% figure is that if the form of the decay curve flattens out significantly (as does happen with some materials) then by the time this level is reached there is not much charge left and little influence on items nearby. The disadvantage is that with some materials this may involve excessively long test times. Decay times should be measured to both levels whenever practical.
- It is suggested that the full form of the decay curve is recorded to be available for any fuller later analysis deemed necessary.
- A convenient way to appreciate how the rate of decay is varying during the progress of decay is to analyse successive parts of the curve as if each were part of an exponential decay with its own local decay time constant value. When this is done with materials showing very long charge decay times it is observed that the local decay time constant comes to increase linearly with time after an initial settling down period. In this situation it is possible to predict the time it will take for achievement of the decay time to 1/e and/or 10% without needed to wait for completion of such times [41].
- In practice materials may be used as supported freely and well away from any earthed surfaces or, at the other extreme, actually resting against an earthed surface. These two extreme situations may be described as 'open backing' and 'earthed backing'. Charge decay measurements need to be made with both these test conditions in assessing general suitability of materials [11,22,26,39,40]. Both open and earthed backing decay times should be reported and the longer value used in assessing the suitability of materials.
- It is considered generally most appropriate to make measurements on test areas that are initially charge neutral, or whose surface voltage is less than say 2% of the expected initial peak voltage. Different decay times may arise if much of the sample surface area is charged, and is decaying at its rate, compared to the behaviour with just a reproducible initial area charged.

- When the plate carrying the corona discharge points is moved quickly out of the way there is some ionization left in the air around the corona points. If this enters the space between the sample surface and the fieldmeter sensing aperture then a reading by the fieldmeter will be generated relating to the space potential in the cloud of ions. The effect of this can be seen if decay time measurements are made on, say, a surface of copper. The decay time of this is typically around 0.1s. A convenient way to minimize this effect is to use an air dam on the trailing edge of the moving plate to sweep the ionized air out of the way. Such an air dam is shown in Annex 2. An additional air dam on the leading edge of the plate would be an added advantage with physical arrangements providing a piston type action to sweep away ionisation.
- With quick removal of the moving plate, for example by tension springs, the impact of the plate when it is stopped may transmit a shock through the instrument. Care is needed to minimize this shock to avoid it interfering with fieldmeter observations by vibration of the fieldmeter chopper assembly.

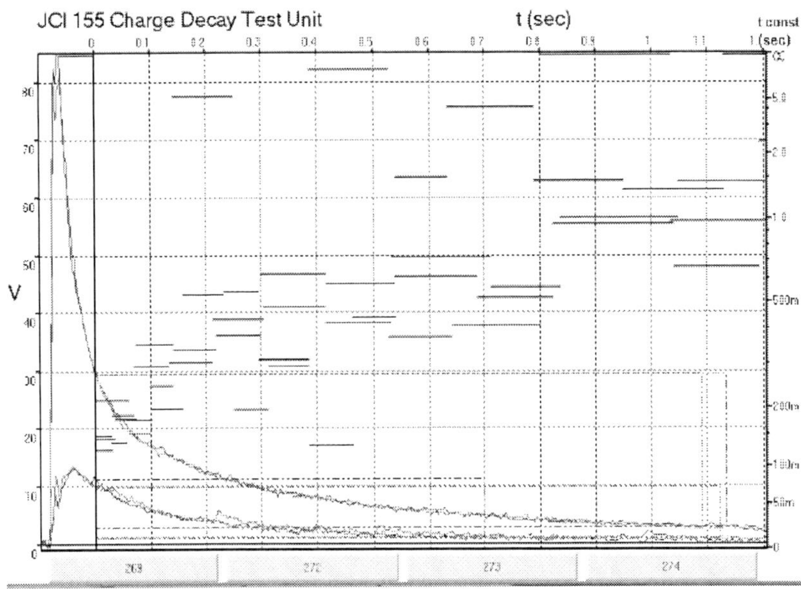

Figure 3.9. Examples of charge decay curves for copy paper with two levels of initial charge and start of analysis 100ms after end of corona charging.

Examples of charge decay curves for copy paper are shown in Figure 3.9 with the start of analysis 100ms after the end of corona charging. The curves show a) reproducibility between repeat tests, and b) independence of decay times to 10% on initial surface voltage.

Procedures appropriate for formal calibration of the surface voltage and decay time performance of corona charge decay measuring instruments are described in Annex 3 [40].

3.4.5. Capacitance Loading

The influence of electrostatic charge on the surface of materials to items nearby depends on the level of surface voltage achieved and on the time for which this is present. If the charge experiences a high ‘capacitance’ on the material then only low surface voltages will arise from the quantities of charge likely to arise in practical tribocharging events. Conversely, if the 'capacitance' is low then high voltages will arise. For instance, fabrics that include conductive threads, such as for personal protective clothing and cleanroom garments, may show high values of ‘capacitance’ where the conductive threads are close spaced. Although some such materials show rather long charge decay times they may present little risk of causing problems if the effective capacitance is high. Thus for many situations it seems that the suitability of materials to avoid problems from retained static charge may be judged by both the charge decay time and by the capacitance exhibited to surface charge [39,40] – see also Annex 2.

It is not possible to derive a proper value of 'capacitance’ from the quantity of charge required to create a measured value of surface voltage. The charge on the surface will not be uniformly distributed so the voltage will be non-uniform over the surface. The area over which charge is deposited (by tribocharging or by corona) is likely to be much smaller than the area of the test aperture over which a uniform potential is used for instrument voltage sensitivity calibration. Hence the peak voltage will be rather higher than the value interpreted from the fieldmeter measurements. Rather than guess an area for the deposited charge, in order to try to calculate a real ‘capacitance’, it is more practical to assess materials in terms of their ‘capacitance loading’ – a ratio of the apparent capacitance of the test material ($C = Q / V_{pk}$) to that provided by a very thin layer of a good dielectric (see Annex 2). It is assumed that there are similar distributions of surface charge for the two materials. This ratio approach takes out concern over non-uniformity in charge distribution and differences between instruments and test arrangements.

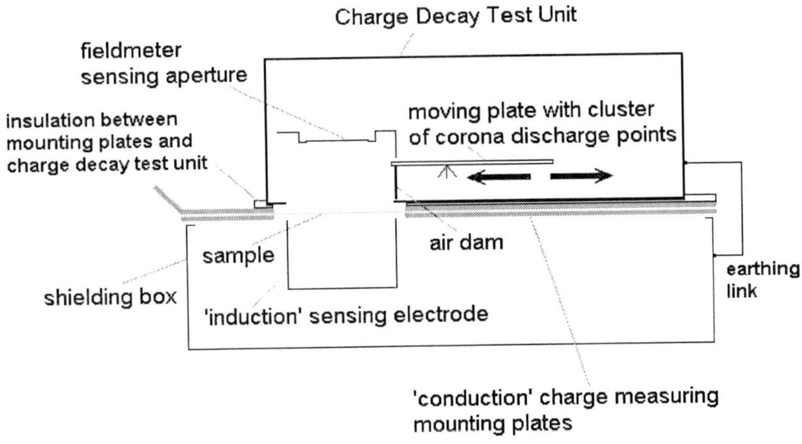

Figure 3.10. Arrangement for measuring received charge.

The charge transferred to a test surface can be measured with the arrangement shown in Figure 3.10. The charge is measured as a combination of the signals observed by the 'conduction electrode' and the 'induction electrode' [17,19,23]. The 'conduction' charge signal relates to the charge that moves or couples fairly immediately to the sample mounting plates. The 'induction' signal relates to charge held for a time near where it has been deposited and from where it is free to couple up to the inside of the charge decay test unit and down to the induction electrode. The induction electrode is conveniently made physically similar to the sensing region of the charge decay test unit so the induction fields from the retained charge is partitioned about equally between the two. The total charge may be expressed as:

$$Q_{tot} = Q_c + f\,Q_I$$

where the factor f is expected to be about 2.

With a simple dissipative sample material, such as paper or cling film, it is observed that nearly all the initial observations are associated with 'induction' charge effects and that this decreases while the conduction signal increases. The total corona charge deposited is of course constant, hence the fall of the induction signal, Q_i, must match the increase in the conduction signal Q_c. Such measurements enable the factor, f, to be derived to give a good constancy between the sum of the two signal variations.

Capacitance loading characteristics of materials vary appreciably with the quantity of charge deposited for some materials [20,39,40]. This is illustrated in Annex 2: Annex F. In such situations it is desirable to make measurements of how capacitance loading (and charge decay times) vary, or not, with the quantities of charge deposited. Charge should be of both polarities over a wide a range of quantities of charge and down to as low levels of charge as practicable. It has been found that the value of capacitance loading extrapolated to zero charge then provides a basis for predicting the surface voltages likely to arise on practical surfaces with tribocharging [22].

If decay times are longer than a second or so then the maximum surface voltage V_{max} (volts) that can be expected in practice for a quantity of transferred charge q (nC) can be calculated from the capacitance loading values extrapolated to zero charge ($CL_{q=0}$) as:

$$V_{max} = n\, q \,/\, (CL_{q=0})$$

where n is a factor (typically around 75) . This relationship arose from studies on cleanroom garments [20]. Practical values for q are likely to be no more than 50nC.

For critical assessment of materials it is important that testing is done under well controlled and Standard values of temperature and humidity with adequate time (for example 24 hours) for accommodation to test environmental conditions [11,26]. This is outlined in Annex 2: Annex E.

3.5. Measurement of Shielding Performance

3.5.1. Introduction

The method of testing the shielding performance of layer materials that has been developed for the ESD Association Standard is based on measuring energy transfer [42]. This is appropriate for the specific area of interest of protecting semiconductor devices and assemblies in transport and storage packaging. However, it does not provide much information for other areas of application or for appreciating how shielding performance is achieved.

The ESDA Test Method for measuring the shielding performance of packaging bags used for microelectronic components and assemblies is illustrated in Figure 3.11. The basis of the test method is to apply a high voltage impulse across the outside of the bag and observe the energy received by a low impedance measurement circuit within the bag. The form of the impulse is what is called a 'human body model' discharge with a current risetime of 2-10ns followed by a 150ns fall time. 'Satisfactory performance' means the energy received within the bag shall be less than or equal to 50nJ for a 1kV discharge. The energy received is derived by integration of the current waveform observed by the current transformer with the 500 ohm load resistor. The instrumentation for this test method requires a high quality current transformer and a high performance digital storage oscilloscope.

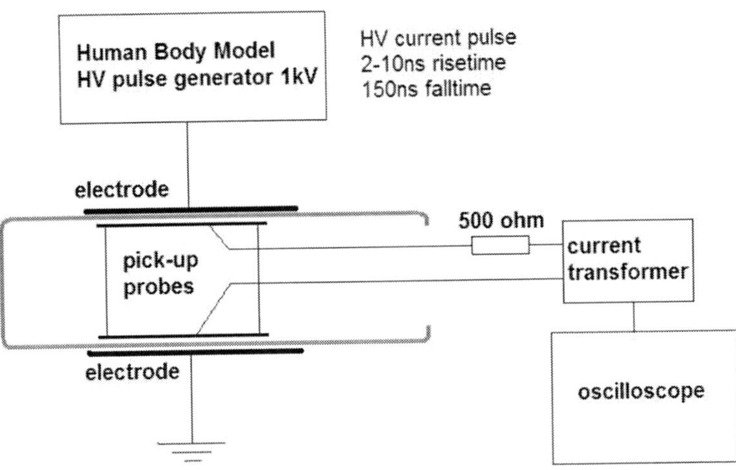

Figure 3.11. ESDA method for measurement of shielding performance.

3.5.2. Alternative Method of Testing

An alternative method of testing has been developed that measures how the electric field shielding performance of materials varies with frequency [42,43,44]. The advantage of the method is that it provides a wider range of information about the characteristics of the material and, in particular, does not require any contact to the resistive features of the layer or any earth bonding contact or connection. Studies with this approach have shown how the performance of some materials, such a 'black bag' films, falls away strongly as the test frequency increases whereas that of films with evaporated metallic coatings varies little with frequency. This suggested the idea that the variation of shielding performance with frequency might provide a unique and non-invasive way to measure the effective resistivity of materials [45].

The approach developed for measuring the shielding performance of film and layer materials [43,44] has five basic features:

a) a plane area of the test material, isolated from earth, is tested by applying an electric field stress across the surface of a defined area of the sample by inductive coupling from a pair of nearby electrodes
b) the electric field stress is applied as a balanced bipolar signal relative to earth. Symmetry about earth potential ensures no common mode signal is impressed on the sample as a whole.
c) measurements are made of the electric field signals inductively coupled to a pair of electrodes on the far side of the sample in positions matching those of the driving pair of electrodes. Observations can be made either as the difference signal between the electrodes on the other side of the sample or from either of these relative to earth
d) shielding is measured at a variety of frequencies as the ratio of signals observed with the material present compared to that without
e) shielding performance is presented as the variation of the ratio of signals with and without the sample as a function of frequency

The basic physical arrangement of the approach is shown in Figure 3.12. In the two test assemblies made to date the electrodes have been 15mm wide and mounted flush in earthed plates to be parallel and with their centre lines 50mm apart. In the first test assembly the electrodes were 50mm long and in a more recent assembly 100mm long. The 100mm length was chosen for studies on the characteristics of fabrics including conductive threads where the threads might be spaced as much as 25mm apart, so a rather wider area of test seemed wise. The arrangement of the observation electrodes matches the electric field stressing electrodes. The inner surfaces of the electrodes and mounting plates have been 10mm apart. The sample is clamped flat against a 3mm thick sheet of good quality dielectric (polycarbonate) on the observation side by a second sheet advanced from the driving electrode side. This arrangement ensures the sample is held flat and in a well defined position relative to the stressing and observation electrodes.

The initial test arrangement [43] aimed to test the shielding performance of materials over the full range of frequencies relevant to risks of damage to microelectronic devices – 10Hz to 1GHz. The stressing electric field was generated by bipolar high voltage pulses of amplitudes up to 10kV. The rise

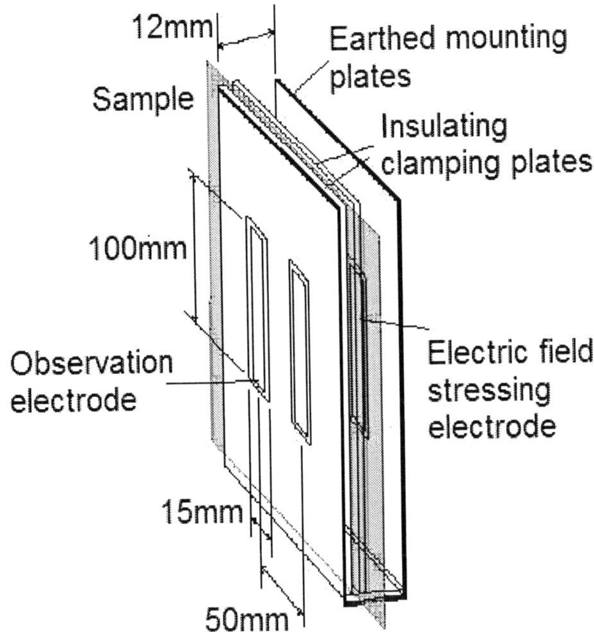

time was around 3ns at 10kV and 1ns at lower pulse voltages. The fall time was arranged to be over 0.1s so the test electric field covered the range of frequencies from 10Hz to 1GHz.

Figure 3.12. General arrangement for shielding measurements.

The signal picked up by the observation electrodes was divided into 9 frequency bands using low Q bandpass filters centered on each decade of frequency from 10Hz to 1GHz. Because the filters needed to respond well to single pulse signals, and narrow bandwidths were not needed, the Q was hence set to unity for all channels except the GHz channel, where it was 0.7. The signals of each filter channel were held in peak detection and hold circuits and these were scanned by a 12 bit analogue to digital converter (ADC) and transferred over a serial data link to a local microcomputer within the time constant of the hold circuits. The computer analysed, stored and displayed observations.

Examples of the variation of shielding performance with frequency are shown in Figures 3.13 and 3.14. Figure 3.13 shows that the shielding performance of a metallized layer type material varies little with frequency, whereas Figure 3.14 shows that the performance of a resistive material (black shielding bag) falls away rapidly with increasing frequency.

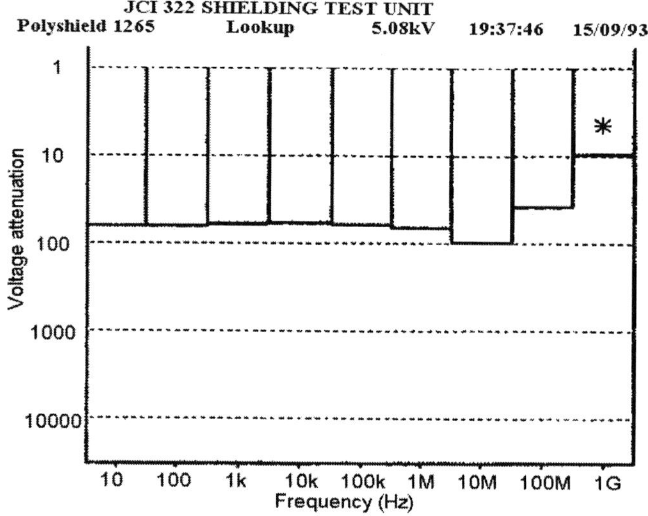

Figure 3.13. Variation of attenuation with frequency for metallised film material.

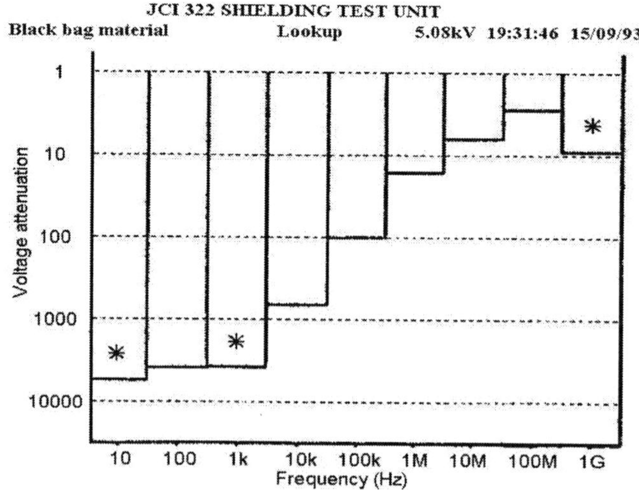

*Figure 3.14. Variation of attenuation with frequency for a carbon loaded 'black bag' resistive material (measurements marked * were somewhat uncertain).*

While the above approach was shown to be basically sound in philosophy the prototype hardware did not operate very stably at the higher frequencies – 100MHz and GHz. Studies on cleanroom garment fabrics, that included conductive threads, showed that attenuation was usually only significant at the lower frequencies. With this in mind a much simpler approach was developed to measure shielding performance on the more limited range of frequencies of 10Hz to 10MHz [44,45]. The same basic arrangement was used for mounting the sample and test electrodes but the stressing electric field was provided as balanced antiphase sinewave signals at a single frequency that was scanned over the frequency range of interest. Because operation is more stable at these lower frequencies it was not necessary to measure the reference signals at the same time as the observations with the test material. The arrangement and basic circuits are shown in Figure 3.15.

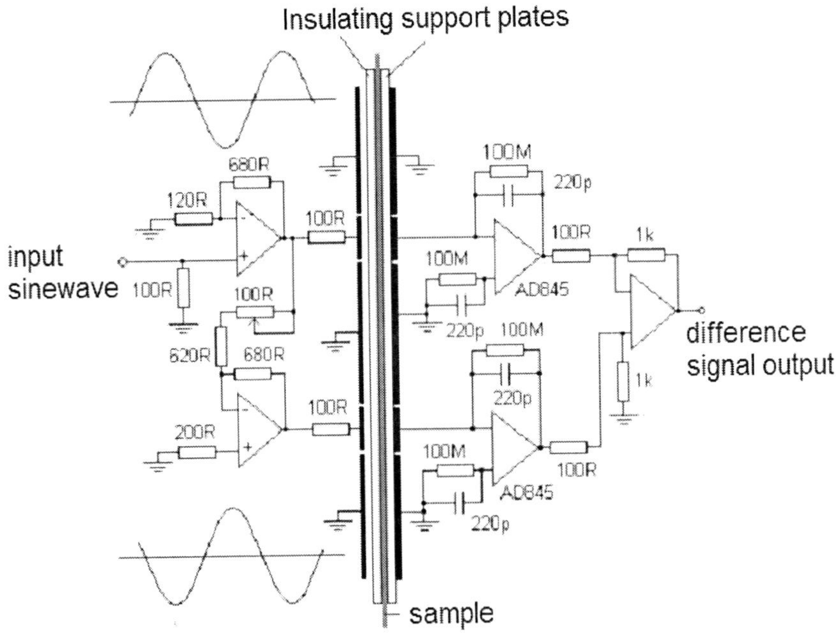

Figure 3.15. Bi-phase sinewave drive circuit and signal difference observation circuit.

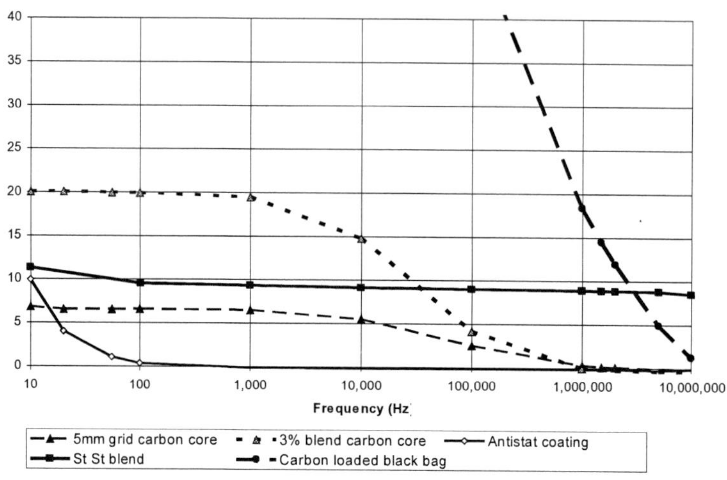

Figure 3.16. Shielding performance of a variety of fabrics and for black 'shielding bag' material

To give good signal to noise ratios at high attenuation the signals observed were phase sensitive detected and filtered for measurement. To take account of the phase shift that occurs as the attenuation rises the signals were also phase sensitive detected in quadrature. This enabled true attenuation values to be derived as the square root of the sum of the squares of the two components.

Figure 3.16 shows the variations of attenuation with frequency for two polyester based cleanroom garment fabrics that included core conductive threads, a fabric with stainless steel fibres blended in, a fabric with just an antistat coating and a piece of carbon loaded 'black bag' film from a shielding bag. These results show that the attenuation of electric field signals by the material that includes metallic threads ('St St blend') is fairly independent of frequency, whereas all others show a decreasing attenuation with increasing frequency. (The results shown in Figure 3.16 were taken before it was recognised that results should include measurement of quadrature components – as are included in the results shown in Figure 3.17. Figure 3.16 nonetheless shows real trends).

To examine the detection sensitivity for very small quantities of conducting thread some measurements were made with samples of single fine copper wires mounted across the interelectrode gap. A nearly uniform attenuation was observed over the frequency range, 10Hz to 10MHz, of around 0.54dB for a single 0.5mm diameter wire and 0.42dB for a 0.2mm wire.

3.5.3. Relationship of shielding to effective resistivity

The variation of shielding attenuation with frequency provides a way to determine the effective resistivity within or on layer materials. It is felt this has relevance to the opportunity to draw incendive spark type discharges from charged materials [46,47]. The effective resistivity may be determined by comparison of the variation of attenuation of the sample with that for a defined value of resistivity [45]. This defined value was created in the arrangement shown in Figure 3.15 by using two 100mm long 15mm wide strips of brass foil mounted on a dielectric sheet between the drive and observation electrodes and linked together by a single resistor of defined value. By way of comparison, measurements were also made with a layer of slightly damp paper whose resistivity could be measured in the traditional way from the resistance between two parallel contact bars.

The variations of attenuation with frequency are shown in Figure 3.17. (The finite attenuation shown by the brass strip and resistor arrangement at high frequencies is probably a result of fringing field coupling that was not present with the continuous area of slightly damp paper).

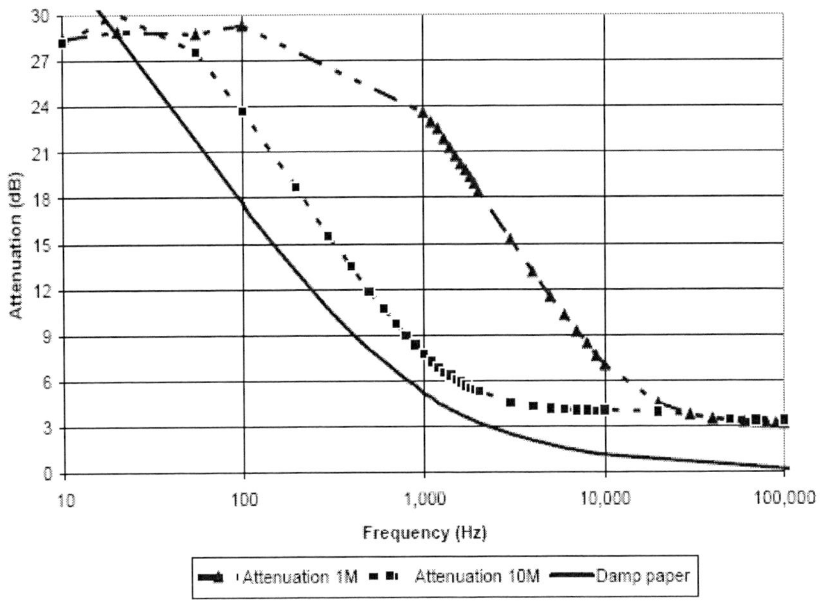

Figure 3.17. Variation of shielding performance with frequency for 1M and 10M resistors between brass foil strips and with damp paper.

The variations of shielding performance observed with 1M and 10M resistors were similar in form but shifted in frequency by a factor of 10. The mid-range variation of attenuation with frequency is about 6dB per octave – which is what would be expected for an RC type filter action. Comparison between these measurements and measurements made with a layer of slightly damp paper (also shown in Figure 3.17) suggested that the effective resistance presented by the 'damp paper' was around 20M. The test area of the damp paper was that between the two 100mm long strips with 50mm separation. The 'resistivity', taking account of the aspect ratio, appeared to be about 40M. Conventional measurement of the 'resistivity' of the damp paper sample with two contact bars gave values 30-50M - which is in at least reasonable agreement [45].

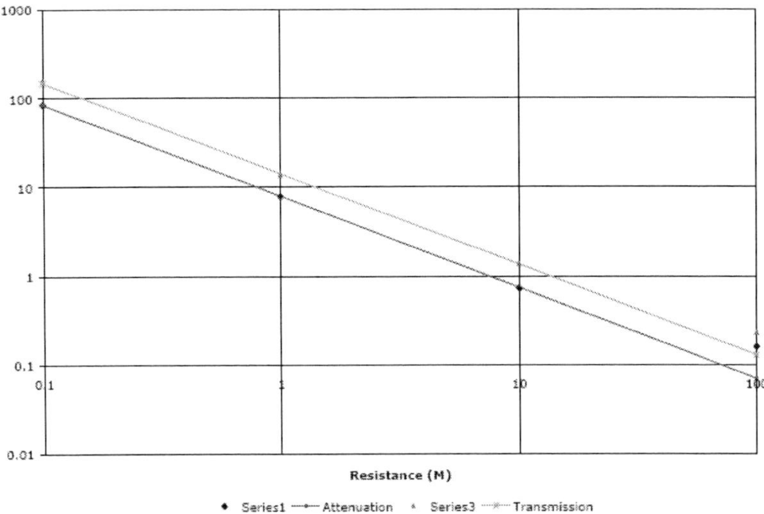

Figure 3.18. Variation of 50% attenuation frequency with resistance.

The variations of frequency (kHz) observed for 50% transmission and for 50% attenuation observed with the resistance value (Megohms) are shown in Figure 3.18. Resistance values (Megohms) may be derived from the frequency f (kHz) for 50% attenuation and for 50% transmission as:

From 50% attenuation fraction: $R = 10^{(0.947 - \log(f))}$

From 50% transmission fraction: $R = 10^{(1.135 - \log(f))}$

To express this as an equivalent resistivity over the test area it is necessary to take into account the ratio of the length of the electrodes to their spacing in the particular test set-up.

3.6. Measurement of the Incendivity of Electrostatic Discharges

There is good discussion of the measurement of the electrostatic spark energy required to cause ignition of flammable gases, vapours etc in the book by Lewis and von Elbe [48] and there are a number of relevant formal Standards available [1,49,50,51,52].

From the instrumentation point of view the features on which special care is required are:

- the spark gap needs to be between spherically ended, clean, smooth and hard metal electrodes to avoid the occurrence of corona at the maximum test voltage used and to minimize pitting with repeated spark discharges
- the high voltage capacitor needs to involve a high quality dielectric so the capacitance value does not change with voltage and there is no charge memory effect. It shall have minimal internal inductance and resistance.
- there must be minimal resistance in the discharge circuit.
- the value of the capacitance must be measured in situ in the test apparatus so that full account is taken of the capacitance of leads and connections. This is particularly important for small values of capacitance.
- the test spark must occur between the test electrodes either by increasing the high voltage supply voltage until breakdown occurs or by applying a defined voltage and closing the gap mechanically. No series triggering spark gap is permissible. With airborne suspensions of powder particles it may be best to use the closing gap approach. This will minimize the deposition of charged airborne particles on the electrodes at a slowly increasing electric field between the electrodes.
- the value of the capacitance should be formally calibrated. The voltage at which the spark discharges occur should be measured with

calibrated instrumentation – either a high voltage resistive divider or an electrostatic voltmeter.
- the concentrations of the gases in the test mixture and the ambient temperature and pressure need to be properly measured and recorded

3.7. Measurement of Charge Transfer and Currents in Discharges

Measurement of the charge transferred and of the current flow in electrostatic discharges from charged surfaces and objects is often made using a quasi-spherically ended probe. The probe may approach the charged surface or the surface voltage may be allowed to build to the breakdown level. Many studies are conducted using unshielded probes and an unshielded probe design has been proposed for measuring charge transfer for a new IEC Standard IEC 61340-4-6 (as for example [53]). It has been argued before [54] that only shielded type probes will enable correct measurements to be made of charge transfer and of current flow. The need to use shielded probes was examined in a recent paper together with demonstration of performance for both conducting and insulating charged surfaces [55]. An example of a shielded probe is shown in Figure 3.19.

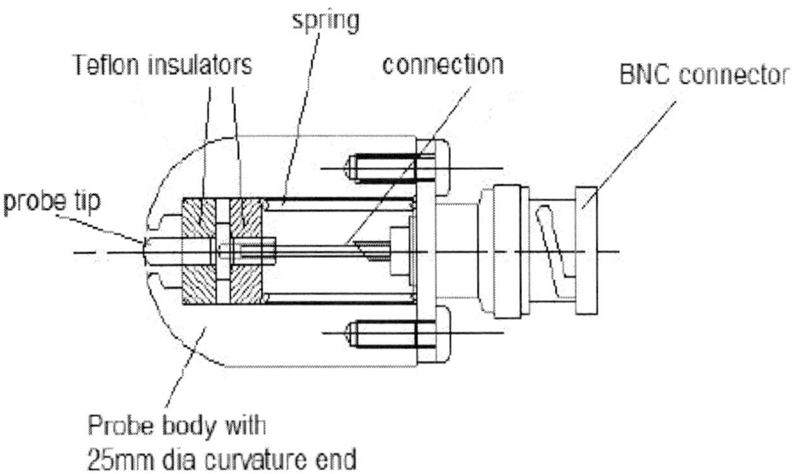

Figure 3.19. Example of shielded probe design.

Measurements of charge transfer and current flow in static discharges may be made by measuring the voltage developed across, respectively, a capacitor or a resistor to earth.

For charge transfer measurements the capacitor needs to be of good quality and of sufficient value to avoid generation of any significant voltage on the probe tip at the maximum value of charge transfer expected (for example less than 20V) to avoid changing the character of the discharge.

If a virtual earth circuit is used for measuring charge transfer in discharges it is important to note that the performance of the circuit is limited by the output drive current capability of the op amp used. If the output drive current is less than the input current then the input may not be held within the supply rail voltages and the output will not match the input. This of course is a particular problem with fast discharges (for example between conducting surfaces) as the amplifier then has to provide both high current and fast feedback. There are two ways round this problem for quantity of charge measurement: first, is to use a buffer capacitor on the input from the discharge probe with a feed resistor that limits the maximum current to within the feedback current capability of the amplifier. This is illustrated in Figure 3.20.

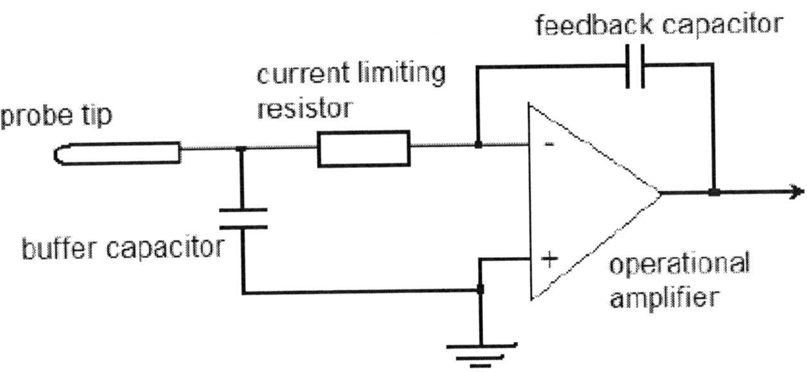

Figure 3.20. Example of buffer capacitor with virtual earth charge measurement amplifier.

The use of an input current flow resistor in a virtual earth charge measurement circuit (as in Figure 3.20) limits the output response time. This is not a problem where just the total charge transfer is to be measured but it is in current flow measurement or where it was desirable to see whether charge transfer was as a single or as close spaced multiple discharges.

The second approach is to measure charge by integrating the flow of current – which might be done by software processing of digitally stored fast observations of current flow.

For fast current flow measurement the resistor to earth needs to be of low inductance and care is needed to match to the transmission line impedance characteristics of the coax cable connection to the display and recording oscilloscope [56]. If a buffer amplifier is used to feed the signal into the coax cable connection the influence of the cable impedance (usually 50R) on the drive amplifier needs to be taken into account.

Spark discharge currents may be up to a few amperes, so it is important that the resistor has a low value (for example 1 ohm) to avoid voltages being developed that may affect the character of the discharge. Static discharges from small items at only a few kV have risetimes down below a nanosecond [57] so great care is needed in measurement arrangements and with the instrumentation used for display and recording of observations to ensure faithful presentation of observations and reliable interpretation.

3.8. Measurement of Resistance

There are plenty of commercial instruments available for measurement of resistance up to around 10^{14} ohms. Instruments are available to measure much higher values but these need to be used with care and appropriate test procedures. Methods for measurement are described in a number of standards [1] and there are a number of calibration houses available to provide formal calibration of measuring instruments traceable to National Standards.

One problem with measuring very high values of resistance is that the reading may change with time after application of the test voltage because the movement of charge is not necessarily linearly dependent on electric field and the distribution of charge varies as it migrates in relation to the capacitance it experiences. This is a particular problem with powders where there is the added question of the packing density to be used. A decision needs to be made at what point to take the relevant reading.

In a number of instances it may be practicable to derive resistance and also current by measuring capacitance and voltage decay rate. An example is measuring the leakage resistance of an electrostatic voltmeter. A charge can be transferred to the input by contact to a voltage source and the resistance then derived from the RC decay time constant (or rate of decay) observed using a measured value for the internal system capacitance. It is not difficult to measure leakage resistance values in this way to over 10^{15} ohms. A special advantage in this approach is that the effective leakage resistance can be measured for voltages up to the maximum operating voltage of the voltmeter and so show up the occurrence of any corona discharges.

3.9. Measurement of Capacitance

There are plenty of commercial instruments available for measurement of capacitance and with the ability to make measurements with a resolution to around 0.1pF. There are a number of calibration houses available to provide formal calibration of measuring instruments traceable to National Standards.

The relevance of capacitance measurement in electrostatic studies is often to assess the capacitance of a person standing on the floor, a container resting the ground or of an item of plant isolated from ground. It may also be necessary to measure the capacitance of electrode structures and low value capacitors involved with ignition testing and the values used within measuring instrumentation.

An important point in measuring low values of capacitance of some pF is to minimise the influence of varying capacitance to the measurement lead. The best approach is to hold the measurement lead via a stand-off insulator to minimise direct hand capacitance. The measurement electrode is then best held just out of contact with the test surface, the 'zero' reading is noted and the reading then taken after the minimum movement to make contact.

References

[1] BS 5958: Part 1: *1991 "Code of practice for Control of undesirable static electricity:Part 1 General considerations"* BS 5958: Part 2: 1991 *"Code of practice for Control of undesirable static electricity: Part 2. Recommendations for particular industrial situations"*

[2] *"Basic specification: Protection of electrostatic sensitive devices. Part 1: General requirements"* EN 100015: 1992

[3] *"Static electricity: Technical and safety aspects"* Shell Safety Committee 1988

[4] *"Electrostatics - Part 5.1: Protection of electronic devices from electrostatic phenomena - General requirements"* CENELEC EN 61340-5-1 May 2001

[5] D. M. Taylor, P. E. Secker *"Industrial electrostatics: Fundamentals and measurements"* Research Studies Press, John Wiley 1994

[6] P. E. Secker *"The design of simple instruments for measurement of charge on insulating surfaces"* J. Electrostatics 1 1975 p27

[7] J. M. van der Weerd *"Electrostatic charge generation during washing of tanks with water sprays, II Measurements and interpretation"* Static Electrification Conference, London, 1971 IoP p 158

[8] R. E. Vosteen *"Electrostatic Instruments"* International Conference on Charged Particles - Management of Electrostatic Hazards and Problems, Oyez 1982

[9] J. N. Chubb *"Two new designs of 'field mill' type fieldmeter not requiring earthing of rotating chopper"* IEEE Trans Ind Appl 26 (6) Nov/Dec 1990 p 1178

[10] J. N. Chubb *"Experience with electrostatic fieldmeter instruments with no earthing of the rotating chopper"* 'Electrostatics 1999' Inst Phys Confr Series 163 p443

[11] *"Methods for measurements in electrostatics"* British Standard BS 7506: Part 1: 1995 Part 2: 1996

[12] J. N. Chubb *"The calibration of electrostatic fieldmeters and the interpretation of their observations"* Electrostatics '87. Inst Phys Confr Series No. 85 1987 p 261

[13] I. E. Pollard; J. N. Chubb *"An instrument to measure electric fields under adverse conditions"* Static Electrification Conference, London, 1975. Inst Phys Confr Series 27 p182

[14] J. N. Chubb; J. Harbour *"A system for the advance warning of lightning"* Proceedings of Electrostatics Society of America Annual Meeting 2000, Brook University, Niagara Falls, Ontario, Canada. June 18-21 2000

[15] J. N. Chubb; J. Harbour *"Operational health monitoring for confidence in long term electric field measurements"* Submitted for publication in J Electrostatics in February 2010

[16] J N Chubb *"Tribocharging studies on inhabited cleanroom garments"* J. Electrostatics 66 2008 p531-537.

[17] J N Chubb *"Measurement of tribo and corona charging features of materials for assessment of risks from static electricity"* Trans IEEE Ind Appl 36 (6) Nov/Dec 2000 p1515-1522

[18] P Holdstock, M J D Dyer, J N Chubb *"Test procedure for predicting surface voltages on inhabited garments"* EOS/ESD Symposium 2003. Las Vegas. 21-25 Sept 2003

[19] J N Chubb *"Experimental comparison of the electrostatic performance of materials with tribocharging and with corona charging"* at: http://www.infostatic.co.uk/cache/Tribo-corona-comparison.pdf

[20] J N Chubb, P Holdstock M Dyer *"Predicting the maximum voltages expected on inhabited cleanroom garments in practical use"* 'Electrostatics 2003' 23-27 March, 2003. IoP Conference Series 178 2004 p131

[21] R Gompf *"Standard test method for evaluating triboelectric charge generation and decay"* NASA Report MMA-1985-79 Rev 2, July 1988

[22] J N Chubb *"Instrumentation and standards for testing static control materials"* IEEE Trans Ind Appl 26 (6) Nov/Dec 1990 p1182

[23] J. N. Chubb *"The assessment of materials by tribo and corona charging and charge decay measurement"* Inst Physics 'Electrostatics 1999' Confr, Cambridge, March 1999 IoP Conf Series 163 p 329-333.

[24] J N Chubb *"Measurement of tribo and corona charging features of materials for assessment"* IEEE Transactions on Industry Applications 36 No 6, Nov/Dec 2000 p1515-1522. (Paper presented at IEEE-IAS Meeting, Phoenix, Arizona 3-7 Oct, 1999).

[25] J N Chubb *"Corona charging of practical materials for charge decay measurements"* J Electrostatics 37 1996 p53

[26] IEC 61340-2-1 *"Measurement methods in electrostatics – Test method to measure the ability of materials and surfaces to dissipate static electric charge"* Feb 2000. International Electrotechnical Commission

[27] US Federal Test Standard 101C Test Method 4046. EIA Interim Standard IS-5-A

[28] J. N. Chubb, P. Malinverni *"Experimental comparison of methods of charge decay measurements for a variety of materials"* EOS/ESD Symposium, Dallas, 1992 p 5A.5.1

[29] G. Baumgartner *"Electrostatic decay measurements"* EOS/ESD Symposium 1995 p5.7.1

[30] *"Protective clothing: electrostatic properties – Part 3: Test methods for measuring charge decay"* Draft prEN 1149-3 (3rd rev) Test method 2 (Triboelectric charging)

[31] R. Gompf, P. Holdstock, J. N. Chubb *"Electrostatic test methods compared"* EOS/ESD Symposium Sept 1999

[32] J. A. Gonzalez, S. A. Rizvi, E. M. Crown, P. R. Smy *"A laboratory protocol to assess the electrostatic propensity of protective clothing systems"* J. Textile Inst 91 (4) 2000 p1

[33] P. Holdstock *"Shirley Method for Charge Decay Time Measurement on a Full Garment. Procedures 1-4"* British Textile Technology Group, Manchester, M20 2RB

[34] J N Chubb *"Tribocharging studies on inhabited cleanroom garments"* J. Electrostatics 66 2008 p531-537.

[35] *"Protective clothing: electrostatic properties – Part 3: Test methods for measuring charge decay"* Draft prEN 1149-3 (3rd rev) Test method 2 (Induction charging)

[36] D. M. Taylor, J. Elias *"An instrument for measuring static dissipation from materials"* J. Electrostatics 19 1987 p53

[37] J. Paasi, T. Kalliohaka, T. Luoma, R. Ilmen, S. Nurmi *"Contact charging method for the measurement of charge decay in electrostatic dissipative materials"* Inst Phys 'Electrostatics 2003' IoP Conf Series 178 p 137-142

[38] J N Chubb *"Comments of methods for charge decay measurement"* J. Electrostatics 62 2004 p73-80

[39] J N Chubb *"A standard proposed for assessing the electrostatic suitability of materials"* J Electrostatics 65 2007 p607-610

[40] J N Chubb *"Test method to assess the electrostatic suitability of materials for retained electrostatic charge"* Document prepared in 2004 for discussion as prospective British Standard. Available at: http://www.infostatic.co.uk/cache/JCITestMethod.pdf

[41] J N Chubb *"Charge decay time prediction"* Short communication to be published in J Electrostatics 2010

[42] ANSI/ESD STM 11.31-2001: Bags

[43] J N Chubb, C Hancox, J Harbour *"Design and performance of new instrumentation to measure the shielding capabilities of materials against electrostatic discharges"* J. Electrostatics 30 1993 p307

[44] J N Chubb *"How effectively do materials shield against transient electric fields"* IEEE Applications Sept/Oct 2002 p13-20

[45] J N Chubb *"Non-contact measurement of the electric field shielding performance and the resistivity of layer materials"* J. Electrostatics 64 2006 p685-689

[46] G. J. Butterworth, E. S. Paul, J. N. Chubb *"A study of the incendivity of electrical discharges between planar resistive electrodes"* 'Electrostatics 1983' IoP Confr Series 66 p185

[47] J N Chubb, P Holdstock *"Risks of ignition from Type D ungrounded FIBC"* J Electrostatics 68 2010 p145-151

[48] Lewis, B. von Elbe, G. *"Combustion, flames and explosion of gases"* Academic Press, New York 1961

[49] IEC 61241-2-3:1994 *"Electrical apparatus for use in the presence of combustible dust - Part 2: Test methods - Section 3: Method for determining minimum ignition energy of dust/air mixtures"*

[50] BS EN 13821:2002 *"Potentially explosive atmospheres. Explosion prevention and protection. Determination of minimum ignition energy of dust/air mixtures"*.

[51] ASTM E2019 - 03(2007) "Standard Test Method for Minimum Ignition Energy of a Dust Cloud in Air".

[52] EN61340-4-4 *"Standard test method for specific applications – Electrostatic classification of flexible intermediate bulk containers (FIBC)"* November 2005

[53] IEC 61340-4-6 and U von Pidoll, E Brzostek H R Froechtenigt IEEE Trans Ind Appl 40 2004

[54] J N Chubb, G.J. Butterworth *"Charge transfer and current flow measurements in electrostatic discharges"* J. Electrostatics 13 1982 p209

[55] J N Chubb *"Measurement of charge transfer in electrostatic discharges"* J. Electrostatics 64 (5) 2006 p301-305

[56] J. Smallwood; G. L. Hearn *"A wide bandwidth probe for electrostatic discharge measurements"* Electrostatics 2003. Inst Phys Confr; Heriot Watt Univ. March 2003.

[57] H M Hyatt *"The resistive phase of an air discharge and the formation of fast risetime ESD pulses"* EOS/ESD Symposium, Dallas Texas 1992 p55

f) It may be that an item happens to be without charge at the time of examination. If an item or material is considered suspect then it is useful to see if it easily becomes charged when rubbed and to see how quickly that charge is able to dissipate. An example is the case of flooring. In this case the fieldmeter may be laid on the floor surface so the sensing aperture is looking across the surface, the metalwork of the fieldmeter rests in contact with the floor surface and the display can be read from above. The floor is rubbed or scuffed, for example with the shoe, just in front of the sensing aperture. The fieldmeter will show if the flooring becomes highly charged and how quickly the charge is able to dissipate.

g) The operation of most fieldmeters is susceptible to water, dust and dirt. Fieldmeters can be built to operate without loss of performance in adverse environmental conditions – but this is not easy [1,2] and the instruments are larger and more expensive. One simple way to improve the immunity of normal fieldmeters to their operating environment is to provide a purge of clean dry air over the electronic circuitry and out through the sensing region to keep the sensing surface mounting and the signal processing circuts clean and dry.

h) Purging a fieldmeter with an inert gas may provide a way to make investigative studies acceptable in situations where flammable gases are or may be present. This, however, needs to be considered carefully and formally agreed with the plant safety officer.

i) It is usually useful to make direct recordings of observations. This shows the time variation of electric field values and allows cross correlation of electrostatic variations to other events (for example features of plant operation). This can be very useful for interpretation of observations and for clear and persuasive presentation of results. A good example is the variation of body voltage during walking on flooring where individual steps are shown by the increase of voltage as each foot is lifted from the floor and the capacitance of the body is reduced. There are several units available today (for example the Picoscope) that quite economically turn a PC or laptop computer into a 2 channel digital storage oscilloscope or datalogger for easy display and recording of observations.

j) Measurements of electric fields can be made with appropriately designed instrumentation immersed in dielectric liquids [3]. Because liquid shear at moving surfaces can cause charge separation it is best to use a fieldmeter geometry and speed of operation that minimizes such effects.

k) It needs to be remembered that the response of a fieldmeter relates to the electric field at its sensing aperture. For calibration purposes the sensing aperture of the fieldmeter is mounted flush with an inner surface of a plane parallel plate calibration arrangement where the electric field at the sensing aperture can be defined directly and uniquely by the spacing of the plates and the voltage difference between the plates (see Annex 3). In most uses of fieldmeters they are held or mounted at some distance from a surface on which there is charge or a potential. The structure and crossection of a fieldmeter is usually modest (some tens of mm) so the electric field from the nearby charged surface will concentrate towards the earthy projection of the fieldmeter. The electric field at the sensing aperture will hence be a complicated relationship between the form of the fieldmeter structure, the size and form of the source surface and the spacing between them. It is then not possible to make a simple calculation of the electric field at the sensing aperture by dividing the potential of the target surface by the separation distance. The most usual requirement for fieldmeter measurements is the potential of a surface. In this application it is best to have the fieldmeter set to provide a direct display of surface voltage at a defined separtion distance – as discussed below.

4.3. Measuring Surface Voltage

If an earthed fieldmeter is near a surface at a voltage then an electric field will be generated at the fieldmeter sensing aperture. The electric field depends in a complex way on the geometry of the fieldmeter and the surface. For operation at a defined separation distance there is a linear relation between voltage and the fieldmeter reading – but not with distance. This is shown in Figure 4.1 below. (Many instrument manufacturers assume or imply a linear relationship - which is not correct!). Fieldmeters may be conveniently set to display the surface voltage for a set separation distance directly. With this arrangement the fieldmeter becomes a 'proximity voltmeter'.

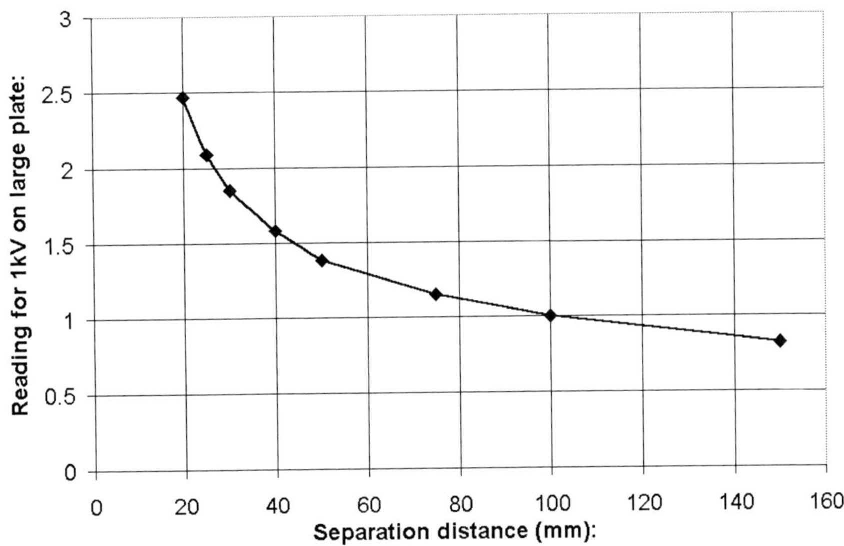

Figure 4.1. Variation of fieldmeter sensitivity with separation distance with reading set for use at 100mm.

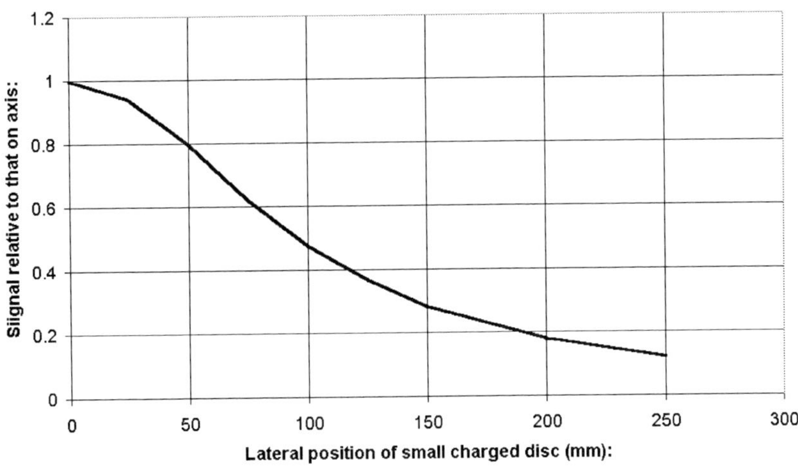

Figure 4.2. Variation of relative sensitivity with lateral position of a small charge.

There are three problem areas with fieldmeter or proximity voltmeter measurements: first, the proximity of the fieldmeter to the surface will add capacitance. This may affect the distribution of charge and the effective surface voltage. This is likely to be a particular problem with charged dielectric layers well away from nearby earthy surfaces. Second, while it is generally good to use a large separation distance to minimize capacitance loading this means that readings will be quite a bit in error for small surfaces. Third, measurements at large separations and will easily be affected by other surfaces and charges nearby. Quantitative measurements may hence require physical or software modelling of the particular practical situation. This is considered later (Section 4.16).

An elegant solution for accurate local surface voltage measurement is the 'voltage follower probe' [4] – as discussed below.

4.4. Electrostatic Voltmeter

An ‘electrostatic voltmeter’ provides the ability to measure the voltage of a conducting item without drawing any charge or current from it. There are three major configuration for an electrostatic voltmeter: first, a mechanically stable mounting of a high voltage electrode near the sensing aperture of a fieldmeter with shielding to prevent any influence by other nearby sources of electric field. The form of the electrode and the form and spacing of shield surfaces must be such as to avoid any risk of corona discharges at the highest voltages to be measured. The second, a pair of specially shaped electrodes which ensure a uniform electric field in the central region where a separate surface is arranged to measure the force arising from the electric field [5]. Third is the ‘voltage follower’, arrangement [4]. This is basically a fieldmeter (usually using a tuning fork or vibrating vane to modulate the source electric field at the sensing surface rather than a rotating chopper) with the voltage of the sensing head unit adjusted to give zero electric field at the sensing surface. In this way the voltage of the head equals the voltage on the nearby surface. This approach has the great advantage that the sensing head adds no distortion or capacitance in the measurement. The probe needs to be mounted close to the surface to give good immunity to influence from any other sources of static charge. This close spacing gives the opportunity for good spatial resolution of charge/voltage patterns. The limitations of this approach are that it is difficult to make instruments that will make measurements on surfaces at many kilovolts and/or where the surface voltage is changing very quickly (probably faster than a kHz).

One particularly useful application of the electrostatic voltmeter, based on the use of an earth referenced fieldmeter, is for the measurement of body voltage. By connecting a person's body to an electrostatic voltmeter with an well insulated lead it becomes easy to monitor the body voltage during such actions as walking over flooring, getting up from a chair and getting out of a car. All these are actions that can create voltages up to 20kV or more with variations in time relating to specific features of body actions. Examples are the increase in voltage associated with the variations in capacitance, for a quasi-constant quantity of charge on the body, when a foot is lifted from and put down on to the floor during walking. It needs to be recognized in such studies that the electrostatic voltmeter will have some internal capacitance (probably only a few pF) and there will be some capacitance added by the high voltage connection lead. The connecting lead will also be susceptibility to the influence of any other nearby varying charges. It is hence best to keep the lead as short as sensible, to avoid it tribocharging by rubbing against surfaces and to avoid varying electric fields in the vicinity.

4.5. MEASURING POTENTIALS IN A VOLUME

An earthed fiedmeter mounted in a region where the voltage was V before the fieldmeter was put there generates an electric field at the sensing aperture of the fieldmeter directly proportional to the difference in potential between the fieldmeter and the previous local potential. If the fieldmeter were at the local potential then the electric field observed would be zero. The electric field at the fieldmeter arises from the way the earthed struture of the fieldmeter locally distorts the distribution of potential – as shown in Figure 4.3. For a cylindrical fieldmeter, of diameter d (m), with its sensing aperture over much of its its end surface, the relation between electric field E and local voltage V is:

$$E = f V / d$$

- where the factor f is near unity [6,7].

This approach is very appropriate for measurement and long term continuous monitoring of atmospheric electric fields. It is also useful in investigation of potential distributions in industrial risk situations. Notable examples have been studies of electrostatic risks during the high pressure jet washing of the cargo tanks of large crude oil tankers [8,9] and electrostatic conditions during filling of large food product silos [9].

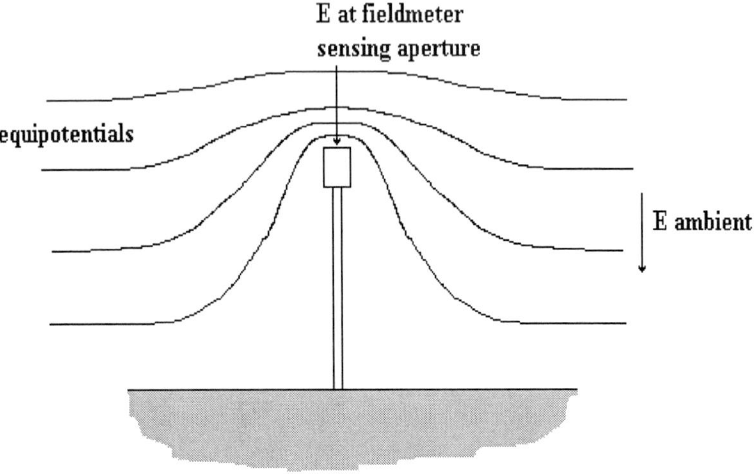

Figure 4.3. Perturbation of potential distribution by earthed projection

The actual sensitivity of measurement of a fieldmeter acting as a potential probe can be checked in the practical situation by applying a calibration voltage to the whole fieldmeter system relative to earth under electrostatically stable conditions. This gives the change in fieldmeter reading as a function of applied voltage. If this is checked for linearity the relationship is established to predict the local voltage from fieldmeter readings. It may be practical to check this prediction by raising the fieldmeter assembly to the voltage at which the electric field reading is zero – when the fieldmeter is at the local potential.

The ambient atmospheric electric field E_a can be conveniently measured using an electrostatic fieldmeter at earth potential mounted on a pole a known distance above ground level. In this arrangement the fieldmeter is used as a probe of the local potential at its mounting height. This mounting arrangement is simple to implement, avoids anxieties about ground level dust, debris and insects entering the fieldmeter sensing aperture and also gives useful enhancement to the basic fieldmeter sensitivity.

The local potential V observed by a fieldmeter of diameter d measuring an electric field of E_{fm} (V m^{-1}) at the fieldmeter sensing aperture is (assuming the factor f above as 1.0) about:

$$V = E_{fm}\ d$$

For an ambient atmospheric electric field, E_v (V m^{-1}) the local voltage at a height h (m) is:

$$V = E_v\ h.$$

Hence the ambient electric field is obtained from measurement of the electric field at the fieldmeter sensing aperture as:

$$E_v = E_{fm}\ d / h$$

There will be a small contribution to the electric field measured dependent on the alignment of the sensing aperture relative to the ambient electric field. If for example the two field components are in directions to add, then the atmospheric field can be expected to be about:

$$E_v = E_{fm} / (h/d + 1)$$

As h/d is normally fairly large the direct influence of the ambient electric field is small.

For operation in wet environments (for example for measurement of atmospheric electric fields) critical gaps need to be at least 6mm to avoid water bridging between plane horizontal surfaces (as discussed in Chapter 3.2.7.6). This requires an appropriately large sensing aperture to achieve sensible coupling and depth of modulation of the external electric field to the primary sensing surface. Insulation for the sensing surfaces needs to be provided with suitably long surface tracking paths and, of course, the signal processing circuits need to be well protected from the environment [1,2]. Use of virtual earth charge measurement input circuits minimises the limitation on leakage to the sensing surfaces. The simplest way to overcome risks of problems from ice and snow is probably to provide the option for some additional heating and to keep the instrument in operation.

Operational health monitoring is important for confidence in long term continuous measurements in adverse environmental conditions (as discussed in Chapter 3.2.7.6). Continuous monitoring of operational health requires creation of an electric field at the fieldmeter sensing aperture that can be reliably separated from the quasi-continuous electric fields being observed [1,2]. This is conveniently achieved by applying an alternating potential either to the fieldmeter assembly or to a nearby shield electrode that is at an odd sub-multiple and phase locked to the chopping frequency. The operational health signal can thus be arranged to provide no interference to the main electric field signal and its strength can be measured by separate phase sensitive detection at its frequency. The value of operational health monitoring was demonstrated some years ago by noting the reduction in sensitivity produced by a spider's web spun across the sensing aperture of a fieldmeter that was monitoring atmospheric electric fields [2].

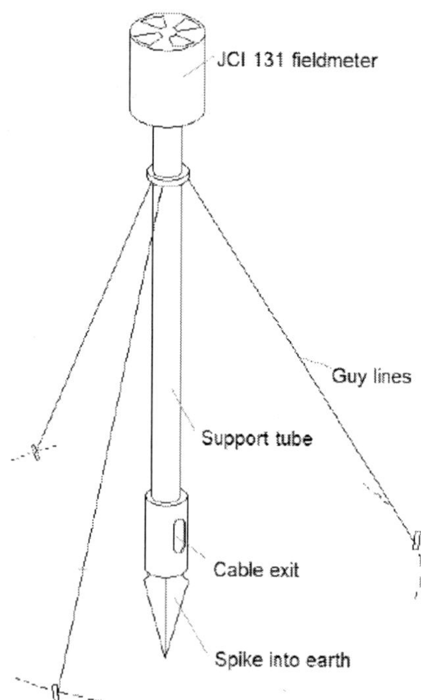

Figure 4.4 .Fieldmeter mounting for measurement of atmospheric electric fields.

It is sometimes necessary to mount the fieldmeter assembly for long term monitoring of atmospheric electric fields in less than electrostatically ideal situations. Examples are the risk of some electrostatic shielding from nearby trees or antenna structures or alternatively the need for the mounting to be in a valley or near the top of a hill. In such situations it is suggested that observations can be normalized by comparison to simultaneous measurements with a hand held or portable fieldmeter mounted over a nearby large flat area of ground under clear blue sky conditions and free of any local sources of atmospheric pollution (e.g. smoking chimneys). The clear and clean blue sky conditions will ensure that the general atmospheric electric fields are the same at the two locations and not influenced by the proximity of clouds.

Space potentials can also be measured using isolated 'water dropper' or radioactive sources. In both cases the potential of the source adjusts itself to the condition that there is no electric field to remove positive or negative charge from the source. At this point the source is at the local potential. The source potential may then be measured by a very high input impedance voltmeter – such as an electrostatic voltmeter. The currents available for charge balancing are very small - so very high quality insulation is needed and the response will be slow. These approaches may be useful for laboratory type experimental studies but are not well suited to practical or industrial measurements – in particular where the humidity is high and/or there may water mist or spray around.

4.6. Measurement of Charge

The Faraday Pail is an appropriate and a very useful instrument for measurement of charge. However, its use needs to be approached with some care. The following points need to be borne in mind:

1) a Faraday Pail only measures the nett charge introduced. So material that contains large quantities of both positive and negative charge may show only a modest nett charge.
2) the charge within the pail and the pail need to be well isolated from the pail surroundings. This means the pail needs to be fairly deep compared to its diameter and to be well shielded from the external environment (see Chapter 3.3.2).
3) If one wishes to measure the charge on samples of a powder it is important to remember that the very act of sampling, for example by scooping up or sliding down a surface, can easily generate additional charge on the powder. The same may well apply if liquid is sampled by flowing along a sampling tube. If the risk is recognized then the influence can be minimized – for example by gentle and careful handling to minimize surface rubbing actions. Where practical it is best for the powder (or liquid) to flow directly from the process into the Faraday Pail without contact to any other surface. Alternatively, it

is probably better to take a relatively large sample rather than a small one - and to avoid tapping the last trace of a powder into the pail. It is particles in contact with the surfaces that will have the strongest electrostatic bonding per unit quantity of material to a triboelectrically different surface.

4) If powder is dispensed into the pail from a spoon or spatula it is best this is of metal and is connected to earth. It is also wise that the operator is bonded to earth and is wearing antistatic clothing. If a plastic surface is used for dispensing then retained charge may create electric fields at the powder separation point and affect the charge carried on the powder. If it is desired to measure charge separation as powder slides down an insulating surface (as a model of a practical problem situation) then it would be wise to start with the insulating surface charge neutralized.
5) If a simple Faraday Pail system is set up using a fieldmeter that happens to be available then it is necessary, of course, to earth the pail to start from a 'zero' reading condition. If such measurements are made in atmospheres that could be flammable this earthing needs to be done with care to avoid any risk of an incendive discharge from the capacitively stored energy if the pail has become charged to a significant voltage. A simple way to avoid this risk is to use a wooden ‘earthing’ contact. This provides a simple high resistance way to dissipate charge sufficiently slowly that no spark discharge will occur.

4.7. SURFACE CHARGE DENSITY

When a fieldmeter mounted in a large plane guard plate is brought up near a uniformly charged dielectric web, nearer than any other earthed surfaces, there is an electric field created:

$$E = (\sigma_1 + \sigma_2)/\varepsilon_o$$

where $(\sigma_1 + \sigma_2)$ is the algebraic sum of the charge densities on the web [5].

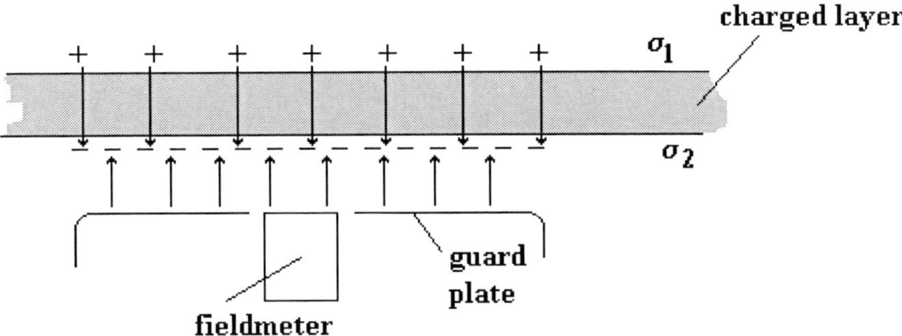

Figure 4.5. Measurement of surface charge density.

If a charged dielectric web is rested against and earthed surface then only the charge on the outer surface is available to couple to a nearby fieldmeter. The electric field, at a guard-plated fieldmeter a distance d (m) away, is reduced by the capacitance effect so that:

$$E = \sigma t / (k \varepsilon_o d)$$

where t is the thickness (m) and k the permittivity of the web. This provides a practical way to measure the charges on each side of web materials.

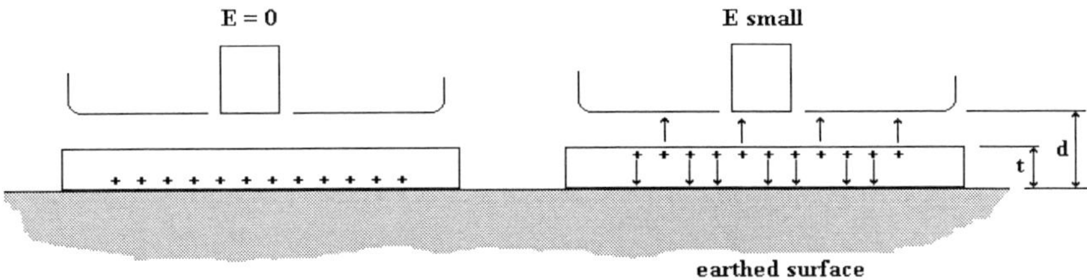

Figure 4.6. Measurements of charge density on a layer against an earthed surface.

For example, a dielectric web of 100μ thickness and permittivity 2 with a charge density of say 10^{-6} C m^{-2} would give a field of about 1.2 10^{5} V m^{-1} freely supported near a guarded fieldmeter and about 550 V m^{-1} if this web were rested against an earthed backing with the charge on the outer surface and the fieldmeter 10mm away. (In terms of a 'capacitance loading' [10,11,12] this would be equivalent to a loading value around 220).

The same sort of suppression of surface voltage may arise from conductive components (layers or threads) buried within materials [10,11,12]. The relevance of this 'capacitance loading' effect can be directly checked by measuring the apparent surface voltage created by a known quantity of charge on the material and comparing this with that for a thin layer of a good dielectric material (see Section 3.4.5).

4.8. VOLUME CHARGE DENSITY

The density of space charge in a volume can be measured from the maximum space potential or from the boundary electric field for simple geometric systems. For example, for a uniformly charged spherical volume:

$$V_{max} = n\, q\, a^2 / (6\, \varepsilon_o)$$

$$E_{boundary} = n\, q\, a / (3\, \varepsilon_o)$$

Charge densities in more complex geometries may be measured using a small scale sampling volume that is transparent to charge movement but providing good electrostatic shielding against external electric fields. This might conveniently be an earthed close spaced wire mesh cube or sphere with a fieldmeter looking inwards from the outer boundary. The electrostatic shielding performance can be checked by raising the shielded volume and the fieldmeter system to a high potential and checking that external field created does not penetrate to significantly affect the fieldmeter zero reading.

The density of space charge may also be determined in complex geometric structures using a fieldmeter that is bonded to earth to plot the distribution of space potential though the volume and then matching this to computer modelling calculations (see 4.16 below). Examples of such situations have been studies in the cargo tanks of large crude oil tankers and in silos [8,9].

Local charge densities may also be measured by drawing a flow of the atmosphere through a suitably fine filter and measuring the volume of air extracted and the current received by the filter.

4.9. Charge Decay for Assessing Materials

The importance of charge decay measurements for assessing the suitability of materials to avoid risks and problems and for the constructive use of static electricity has been discussed in Chapter 3.4. This emphasised the importance of using a correct test instrumentation and test procedure (Annex 2 outlines proposed approaches). These need to give demonstrated matching of characteristics to those observed with tribocharging with a variety of sample materials [13]. In particular the suitability of corona charging is noted as a basis for compact and easy to use instrumentation that satisfied these requirements.

Some points that need to be noted in making charge decay measurements:

1) With thin fabric, film and layer materials it is important to stretch the test area to be flat. This avoids the sample surface being rubbed by movement of the air dam used to remove residual corona air ionization. It also minimizes the risk of oscillatory signals arising in the observed values of surface voltages due to flapping of the surface of the material. For light powders with long decay times it is wise to minimise disturbance of the sample surface by making measurements with a slow release of the plate carrying the corona discharge points.
2) With thin fabrics and with film and layer materials it is generally desirable to make measurements both with an open backing and against an earthed backing surface. These conditions represent the extremes of practical application. The longer of the decay time values observed should be used to judge acceptability of the material.
3) Where an appreciable surface voltage (more than say 5% of the expected initial corona charging peak voltage) is observed after the sample is mounted into the test position then decisions need to be made as to how best to proceed. If the voltage is falling away moderately quickly then it will be best to wait for this to fall to a suitably low level before carrying out a test. If the observed self-decay is quite slow then it may be useful to start by measuring the self-decay of charge on the surface. Self-decay characteristics may not match closely to the decay characteristics observed with corona charging because the surface distribution of charge may be very different from the usual localised patch of corona deposited charge. The charge distribution pattern will be more similar to that at a later time in

corona charge decay when the initial local patch of charge has spread out over the surface and the rate of decay has slowed up.

4) Long charge decay time materials easily become charged to appreciable levels in normal handling. In such situations simple charge decay tests may usefully be made just using a stable support for the samples and a fieldmeter to monitor the self-decay. As well as mechanical stability it is important to provide shielding against any other sources of charge around – including people! With a defined volume of powder, or liquid, the results should be acceptable for comparison of the performance between different materials.
5) Experience with a number of liquids [14] has shown reasonable matching between decay time and volume resistivity according to the textbook relationship, $\tau = e\ e_0\ \rho$. With some liquids the decay follows a nearly exponential form but with others the rate of decay slows up quite a bit during the progress of decay – as is observed with most solid material surfaces. While conductivity measurements may be more convenient with liquids it is recomended that matching to decay peformance should not be assumed.
6) Charge decay times with many powders are very long – for instance with many pharmaceutical powders and paint powders. In such cases the comments above need to be noted. Where decay times are likely to be over say 10,000s it is not usually practical to make measurements out to the 1/e level. In such situations it is useful to analyse the voltage decay curve to show the variation of the rate of decay during the progress of decay. This is best done by calculating the local decay time constant for small steps over the voltage decay curve as if the form of the decay curve over these short intervals did follow an exponential. When this is done two opportunities become available:
    a) it will be noted that after an initial period the local decay time constant increases rougly linearly with time. This provides an opportunity to calculate decay time to 1/e, to 10% or to some other end point, for comparison between materials [15]. For such predictions to be useful the surface voltages of the decay curve need to be measured with a resolution and stability better than 0.1% and with the advantage of 'stutter timing'.
    b) a simple comparison between materials may be made in terms of the local decay time constant at a time of, say, 1000s

7) when making charge decay studies on film or layer materials it is useful to measure the quantity of charge transferred [16,17] (see also Annex 2). Calculations can then be made of the 'capacitance loading'. If this is done over a range of quantities of charge it will be noted that there is a roughly linear variation of capacitance loading with quantity of charge. The intercept of the positive and negative corona measurements at zero charge then provide a basis for calculating the maximum surface voltages that are expected on the material as a function of the quantity of charge placed on the material [18,19].
8) In many situations a high capacitance loading is an adequate alternative to a short charge decay time – but some charge decay is needed to avoid progressive build up of charge with repeated charging actions. An example where a high capacitance loading was not appropriate for overcoming an electrostatic problem was with thin steel sheets with a protective coating on one side for drinks cans that were sticking together in a stack during manufacture. When many sheets had been collected into a stack it was difficult to separate them and this caused problems with subsequent sheet handling. Fieldmeter measurements showed very low surface voltages when the coatings were rubbed, in the range 100-200V. This voltage was stable – so the decay time was long. Because the protective film was thin and on a metal sheet backing the capacitance loading was obviously very high. On consideration it was realized that as the sheets were very smooth and flat so the metal side of one sheet would come closer to charge on the plastic film than the thickness of the film. In this situation large electric fields would arise between the surface charge on the coating and the overlying metal surface. This was able to exert the high forces observed clamping the sheets together. This situation may also apply for dust settling on surfaces where charge may be retained on a thin surface coating and particles held by short range electrostatic attraction. From these examples one may conclude that high capacitance loading is only useful to control problems arising from the influence of surface charge at some distance from a surface – but not necessarily on a surface. Such problems will be overcome by ensuring the surface can dissipate charge, in the high capacitance loading situation of its application, with a decay time of no more than several seconds. This is also noted in Annex 2.
9) The charge decay time characteristics of many materials varies appreciably with humidity. Standard test conditions, to allow fair

comparison between materials, require that measurements be made at the defined levels of 23C and 50%RH and 23C and 15%RH [20,21 and Annex 2]. Other defined levels may be more appropriate for particular industries or applications. The surfaces of materials need to be exposed to these conditions for at least 24 hours to ensure full accommodation to the defined conditions. Measurements must also be made directly within these conditions. It needs to be recognized that any elevation of temperature within measuring instrumentation will affect the ambient humidity there – and 1C will be responsible for a local change of about 2% in RH. It is hence desirable to minimize heat dissipation within charge decay measuring instruments and to measure the internal temperature and humidity.

4.10. Measurements on Discharges

Care need to be taken in measuring the quantities of charge transferred and currents in electrostatic discharges [22,23]. As illustrated in the following diagrams it is important to use a shielded probe and ensure that the discharge takes place to the probe tip. The same approach needs to be adopted for discharges to surfaces of insulating and composite materials [23]. The validity of this has been confirmed experimentally [23].

Figure 4.7 (a) shows a plausible electric field distribution associated with a charge sphere near a large metal plate. Figure 4.7 (b) shows that at discharge only the fraction of the sphere charge that was coupled to the outside world, rather than the plate, is observed by the measurement circuit. Figure 4.7 (c) shows that with a shielded probe all the charge transfer associated with the discharge flows through the probe and into the measurement circuit. Figure 4.8 shows an example of a practical probe design.

Electrostatic discharges can involve current risetimes down to below 1ns [24]. Proper measurements of current hence require the use of very fast response recording oscilloscopes with careful impedance matching of observation signals into and out of the coaxial cable to avoid risk of reflection effects. This means that the resistance Z (Figure 4.7 above) must have a very low inductance and resistor R needs to be equal to the characteristic impedance of the coaxial cable, which needs to be matched at the input of the oscilloscope.

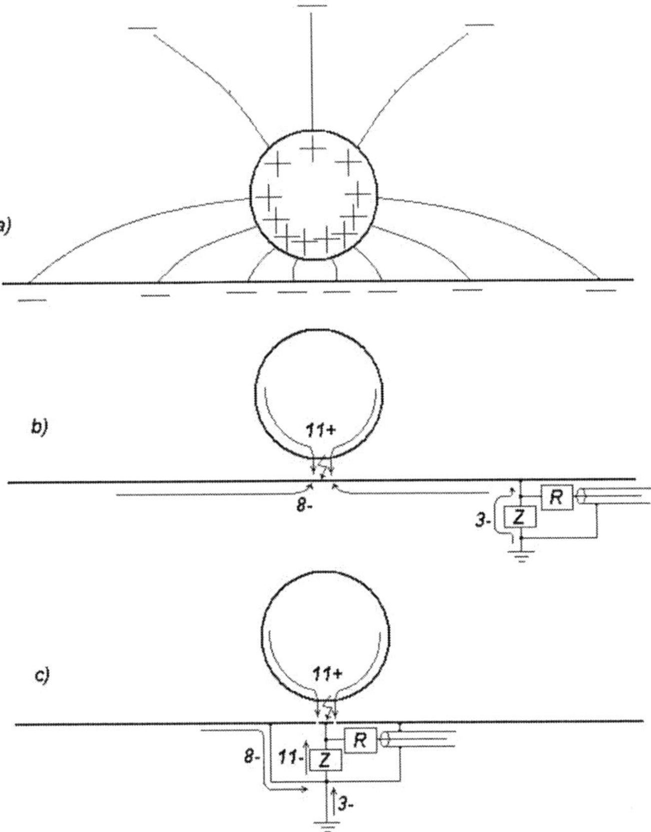

Figure 4.7. Arrangements for measuring charge transfer and currents in electrostatic discharges.

To avoid risk of the discharge spreading from the probe tip to the surrounding shielding surface the impedance Z needs to be kept low. For current flow measurements, a value of 1R is usually appropriate. This may conveniently be constructed from say ten 10R surface mount resistors in parallel arranged to minimise stray inductance.

If a virtual earth circuit is used for charge measurement then it will be necessary to have an input buffer capacitor suitable to absorb the initial charge inflow with an input resistor to the virtual earth amplifier that limits the charge neutralising current to within the current capability of the amplifier. A similar need would arise if a virtual earth amplifier were used to measure discharge currents.

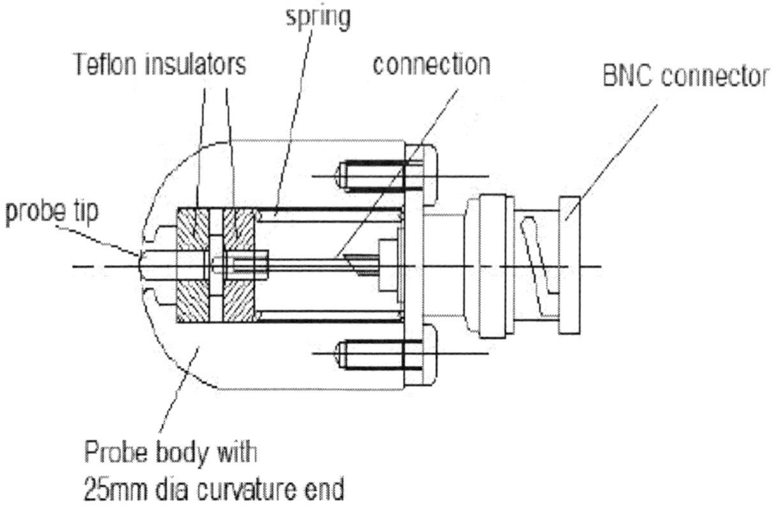

Figure 4.8. Example of design for a shielded probe.

It is important to appreciate features in practical situations that may be relevant to the occurrence of discharges so that measurements relate to practical conditions. The speed of approach may be very relevant to the character and energy of the discharge that occurs when an earthed projection approaches a charged surface. This situation arose in consideration of discharges to the surfaces of water slugs during studies on explosions in large crude oil tankers during tank washing operations [8]. If the speed of approach is very slow the water surface will disrupt with a low energy discharge. On the other hand if the speed of approach is too fast then even if the energy released in the resulting spark discharge is above the minimum ignition energy for the flammable gas mixture present the nascent flame kernel can be squashed out before it has time to propagate into the surrounding gas mixture [8]. With charged dielectric surfaces, if the surface charge can migrate towards the region of high electric field as the earthed projection approaches the resulting discharge will be affected by the speed of approach relative to the speed of charge migration. If there are localized projections on either surface then a corona discharge may occur as the surfaces approach. This may dissipate charge in a safe manner and, depending on the speed of approach, may affect the energy available for the discharge.

4.11. RADIO DETECTION OF SPARKS – RADIO SPECTRUM

The occurrence of spark type electrostatic discharges can be monitored fairly easily and at high sensitivity using radio detection circuits [25]. Radio emission depends on the rate of change of current in a discharge, so spark type discharges radiate much more effectively at high frequencies than do corona type discharges. Sparks may be incendive, if the discharged energy is sufficiently high, but corona is generally considered to be unable to cause ignition. Thus radio detection provides a good way to monitor the occurrence of the type of electrostatic discharge events that could cause ignition and to relate these to relevant operational activities.

Radio detection is effective over a wide range of frequencies - and 38MHz is a convenient observation frequency as it not within broadcast bands [9,26]. The radio receiver needs to be based on an r.f. amplifier stage followed by signal detection. A superhet configuration is not suitable because differences in phasing between the incoming signal and the local oscillator can give wide variations of signal output. The antenna can be quite simple and may be a simple dipole or a tuned loop with coupled pick-up loop to match coupling to a connecting coax cable impedance. What is important about the antenna is to shield it in insulation with a sizeable radius of curvature and/or mount it in a region of low electric field to minimize risks of influence by surface corona or water spray discharges that will be close coupled to the detection circuits. The influence of such discharges close to an antenna can also be minimized by using two separate radio detection units with a coincidence circuit to respond only to events coincident between the two antennae spaced well apart. If observations are to be made within a conducting enclosure than radio observation of the external environment in anti-coincidence will improve confidence in internal measurements. Radio detection observations have been used in a number of shipboard studies into the occurrence of electrostatic sparks during the washing of the cargo tanks of very large crude oil tankers with high pressure water jets and to relate the occurrence of spark type discharges to features of washing operations [8,9].

The radio emission spectrum can provide useful information on the character of spark discharges and help identify the structure and position of sparks [26]. However, as spark events are individual it would be necessary to have a single shot spectrum analyzer to get useful information from such observations!

Studies in the microelectronics industry have shown it is feasible to use radio observations to locate the source of sparks in three dimensions by measuring the relative time of arrival of initial signal edges at a number of antennae spaced around a work area to allow triangulation of observations [27,28].

4.12. Lightning Warning

Clouds in the atmosphere contain charged water droplets. In thunderstorm conditions the electric fields within clouds become sufficiently high that long high voltage lightning discharges can be initiated. If the electric field between the cloud and ground is sufficiently high then lightning discharges can propagate down to earth. For this propagation to occur the atmospheric electric field at ground level under a thundercloud will be 5kV m^{-1} or more.

Advance warning of the risk of local occurrence of lightning discharges is important in a variety of industries – e.g. during electrical connection of mining explosive fuses, in loading military munitions and for personnel working in exposed outdoor situations.

A number of lightning warning systems are commercially available. Measuring the time of travel of low frequency lightning impulse signals to 3 or more base stations enables the location of existing lightning activity to be established and tracked. For assessing the risk of local occurrence of lightning some systems rely purely on measurement of local atmospheric electric fields while some others combine local atmospheric electric field measurement with assessment of the size of electric field steps associated with the occurrence of lightning events in the locality.

An old Nitro Nobel system assessed the risk of local occurrence of lightning by combination of measured values of local atmospheric electric field with observations of lightning events from radio signals at 2-200kHz and radio noise in a narrow band at 27kHz. An update of this system has been developed (JCI 504). This was based on use of an adverse conditions electrostatic fieldmeter (JCI 131) for measurement of local atmospheric electric fields. The mounting pole for the fieldmeter was used as the antenna for radio observations using circuits included within the casing of the fieldmeter. Operational health facilities were included for the electric field and for the two channels of radio observations. These operational health facilities provided confidence in long term continuous monitoring in even very adverse weather conditions [2]. The risk of local lightning was assessed via software comparison of values of electric field, the strength of noise signals and the rate of occurrence of lightning impulse signals against set threshold signal levels.

The occurrence of a high local atmospheric electric field (measured with proper account of any local shielding or enhancement effects as noted in 4.5 above) may not provide much advance warning of a high risk. This is because the high field region below the lower cloud of a multiple charge centres of thunderclouds may not extend very far laterally. There is also the problem that cloud structures and charge levels are likely to vary appreciably from storm to storm. Combination of the local electric field with observations of pre-existing lightning activity broadens the range for advance warning but cannot provide definite warning times or distance because of the differences between thunderclouds.

4.13. Incendivity of Electrostatic Discharges

The risk of ignition of flammable gases and powders by a capacitive electrostatic discharge relates primarily to the energy dissipated in the discharge [29]. There are influences from the duration of the discharge, from the radius of curvature of the electrodes, from the size of and the electrode gap, the temperature, pressure and oxygen concentration [29].

For discharges between metal electrodes the energy U (J) in the discharge will be:

$$U = \frac{1}{2} C V^2$$

where C is the capacitance (farads) and V is the voltage (volts). There is no accepted way (yet) to determine the effective energy released in discharges to isolated dielectric and composite materials. With dielectric layers resting on an earthed backing the incendive risk from 'propagating brush discharge' has been fairly well characterized [30,31,32].

In many practical situations the capacitance on which the discharge energy is stored is associated with some finite conducting body – a person, a metal drum resting on the ground, etc. In such situations the inductance in the discharge circuit in minimal and the resistance is that associated with the nature of the body involved. In building test equipment to measure minimum ignition energies care must be taken in the selection of capacitors and circuit arrangement to ensure there will be minimum inductance (as noted in Chapter 3.6). System resistance should be low as the incendivity of discharges may be affected by the time duration of the discharge in relation to the flame kernel formation time. It is also important to ensure that the breakdown gap is sufficiently large. Powders present their own problems. They are usually dispersed into the discharge gap just before occurrence of the discharge. However it seems that ignition probability may be affected by dust residues from preceding tests and by whether an ignition has just occurred.

Ignition is a statistical occurrence, so measuring the incendivity of electrostatic discharges is difficult. This means that many tests are needed under particular conditions to determine a probability of ignition. Simultaneous observation of other relevant features of discharges should help understanding of ignition probability – for instance from observations of charge transfer, discharge current, low light level photography, sound output and the size of the flame kernel created (via, for example, using shadowgraph or Schleiren photography).

Studies have been made on the incendivity of discharges between an earthed probe and charged insulating layers [31,32,33,34], to insulating layers on an earthed backing surface [31], to liquid surfaces [35] and between an earthed probe and charged clouds [36,37].

One problem that needs to be noted with incendivity tests to liquids is that the surface is likely to be unstable in the high electric fields that will arise just before breakdown – as noted in 4.10 above. With rough surface materials, such as fabrics (particularly those including conductive threads), there may be opportunity for corona discharges before spark breakdown. This would dissipate charge and reduce the local electric field and discharge energy values. In both cases these effects need to be counteracted by using an appropriately fast speed of approach between the discharging surfaces [8]. In general it is wise to try to measure the presence of any pre-spark discharge currents or charge flow.

Where discharge surfaces are dissipative (relatively short charge decay time) charge could migrate towards the high field region before occurrence of a discharge. This should also be guarded against by choosing a speed of approach appropriate to the decay time characteristics of the surface.

The incendivity of discharges to materials seems likely to be affected by the resistivity accessible to a spark discharge [38]. Materials such as cleanroom garments with embedded conductive threads (surface or core conductive) and packaging materials with inner metallised layers are examples where the resistivity accessible to a static discharge could be very different from that measured as a surface property. The accessible resistivity may be the explanation for the ignitions observed on some materials when they are at high humidities [39]. Recent studies with fabrics including conductive threads (Type D FIBC fabrics) showed that avoidance of incendive discharges required the effective area resistivity needed to be no more than 2M [40]. A lower limit was not examined – but will exist.

Testing incendivity with tribo or corona charging has limitations. The surface voltage of sample materials tested may be limited by lack of charge transfer and/or by surface charge migration and by the charge experiencing a high capacitance. The actual surface voltage achieved should hence be measured immediately prior to the time of occurrence of the test discharge.

4.14. Shielding

Shielding against electric field transients is needed to protect sensitive microelectronic devices and assemblies against electrostatic damage while in electrostatically unprotected areas. Body voltages can rise to around 20kV and spark discharges could arise at any voltage up to this when, for example, hand carried packages contacted earth or a package was contacted by the hand or put down on to a metal surface. Current risetimes may be down to 1ns, or less [24]. As device sensitivities may be down around 100V or less, it can be appreciated that good shielding performance needs to offer electric field attenuations of at least 200:1 over the frequency range from 10Hz to 1GHz. Methods of performance measurement need to be able to cover this range.

Shielding performance depends upon conductivity within the material. The resistivity accessible to spark discharges is also relevant to the opportunity to draw energetic electrostatic sparks from charged surfaces [34]. Studies of shielding performance can hence be expected to have relevance to risks of incendive discharges [36,40].

From the device protection point of view, the most desirable method to measure shielding performance would be to have a fully isolated discharge source or detector inside the shielded enclosure and the corresponding item outside. This raises many practical difficulties. Probably the nearest approach would be a detector inside the enclosure that modelled the type of device to be protected by the enclosure but included elements that would indicate exposure to electric fields and currents of the level of interest with balanced sensitivity over the frequency range of interest. The nearest approach might be a semiconductor device with a number of breakdown gaps and fusible elements of known failure characteristic.

Present methods for assessing shielding performance use application of a unipolar, short risetime electric field pulse with differential voltage or current transformer oscilloscope observation of signals to a 500R load. A 'human body model' type discharge pulse is used with performance assessed in terms of the fractional energy transfer through the sample [41]. A deficiency of present methods is they provide no information on the variation of performance with frequency so no guidance is given on suitability of materials for waveforms other than the test waveform. An approach was developed to provide this information, as described in Chapter 3.5 [42,43] but his has not been developed into commercially available equipment.

The approach developed by STFI for assessing materials [44] seems to provide information on the opportunity for occurrence of incendive sparks from charged material surfaces. However, studies on Type D FIBC materials [40] indicate that the STFI method only works where there is earthing of the conductive threads – as for Type C FIBC.

4.15. Other Measurements

Methods are described in the literature for measurements of low currents and resistance/resistivity. Capacitance is another basic parameter often needing measurement in practical situations. Other aspects of measurements are considered in relation to the characteristics of materials in Chapter 3.4.

In making measurements of high values of resistance and low currents it is important to recognise the risks of leakage currents. These may arise over the surfaces of connecting leads and mounting jigs. Guard rings are a useful way to protect against stray current flows.

When measuring low values of capacitance with connections made via flexible leads it is important to isolate the measurement lead from capacitance of the tester's body (for example using an insulated handle, such as a screwdriver) and to make measurements as the difference of readings just out of contact and in contact – as noted in Chapter 3.9.

4.16. MODELLING

With studies on practical plant it is often not feasible to make measurements in the preferred location from an electrostatic point of view. Modelling by physical, analytic [45] or computer modelling [46,47,48] provides ways to relate observations at places which are accessible to measurement to values of potential, electric field and charge density at other places in a system.

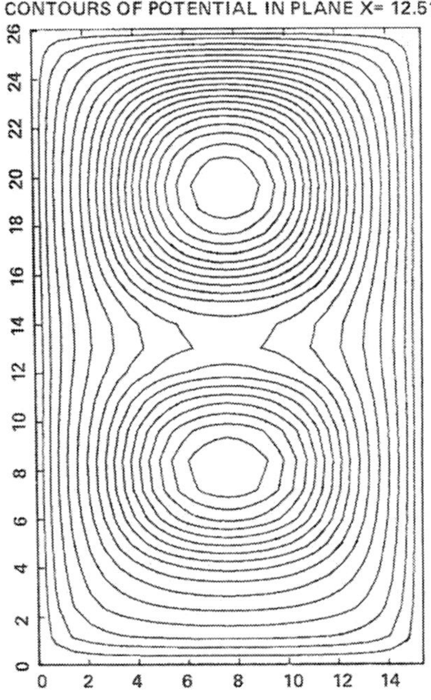

Figure 4.9. Computed potential distribution through the fieldmeter position in tank 4P (for charge density 10nC m^{-3}) 9.5kV max potential, 0.5kV per contour Comparison of computed and fieldmeter measured variations of potential

It is often necessary to make measurements of surface voltages on small items and/or in situations where observations are likely to be influenced by nearby surfaces. One practical approach is to substitute the item by an isolated conducting item of the same dimensions with an electrostatically shielded lead to an external voltage supply. Application of defined voltages to the conducting model item will produce a relationship between voltage and the reading of the fieldmeter.

Computer modelling can handle quite complex structural arrangements, and in two and three dimensions. However, it would be very unwise to base interpretation of conditions of a complex situation on electrostatic observations at any single location. Modelling may well involve some assumptions about voltages on surfaces and/or the uniformity of space charge and/or the fraction of volumes filled with charge. It is wise therefore to use multipoint fieldmeter observations and/or to explore the variation of voltage through some part of the volume to examine such questions. This provides a way to check the predictions of modelling against practical experience [8,9]. Figures 4.9 and 4.10 show examples of potential distributions calculated in large crude oil tanker cargo tanks during tank washing – with Figure 4.10 showing correlation between computed and measured values.

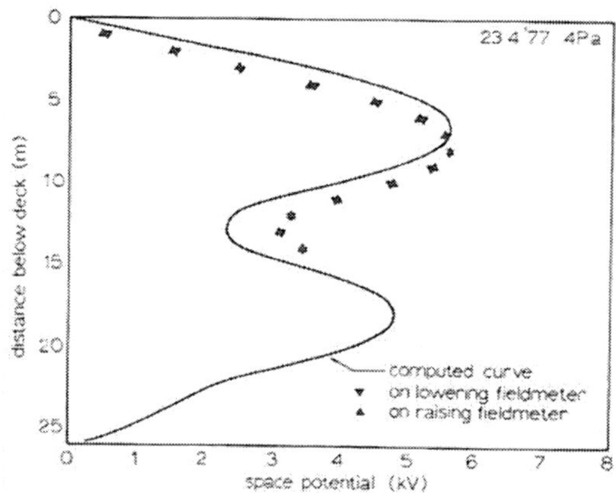

Figure 4.10. Comparison of computed and fieldmeter measured variations of potential

4.17. Calibration

To give confidence in the results of measurements, to satisfy ISO 9000 and to support any contractural or legal requirements, it is necessary that electrostatic measuring instruments are formally calibrated. Suitable methods for a number of basic measurements are described in British Standard BS7506: Part 2: 1996 [20] and, updated, in Annex 3. Formal calibration needs to be made using instruments whose measurement accuracy is traceable to National Standards. In many situations it may not necessary that electrostatic measurements are made to high accuracy - but there is need for confidence in the values obtained to be within known levels of accuracy.

REFERENCES

[1] I. E. Pollard; J. N. Chubb *"An instrument to measure electric fields under adverse conditions"* Static Electrification Conference, London, 1975. Inst .Phys. Confr. Series 27 p182

[2] J. N. Chubb; J. Harbour *"A system for the advance warning of lightning"* Proceedings of Electrostatics Society of America Annual Meeting 2000, Brook University, Niagara Falls, Ontario, Canada. June 18-21 2000

[3] J N Chubb *"Developments in electrostatic fieldmeter instrumentation"* J. Electrostatics 14 1983 p349-358

[4] R. E. Vosteen *"Electrostatic Instruments"* International Conference on Charged Particles - Management of Electrostatic Hazards and Problems, Oyez 1982

[5] D. M. Taylor, P. E. Secker *"Industrial electrostatics: Fundamentals and measurements"* Research Studies Press, John Wiley 1994

[6] J. M. van der Weerd *"Electrostatic charge generation during washing of tanks with water sprays, II Measurements and interpretation"* Static Electrification Conference, London, 1971 IoP p 158

[7] J. N. Chubb *"The calibration of electrostatic fieldmeters and the interpretation of their observations"* Electrostatics '87. Inst. Phys. Confr. Series No. 85 1987 p 261

[8] J.N Chubb *"Practical and computer assessments of ignition hazards during tank washing and during wave action in part ballasted OBO cargo tanks"* J. Electrostatics 1 1975 p61.

[9] J N Chubb, G J Butterworth *"Instrumentation and techniques for monitoring and assessing electrostatic ignition hazards"* Electrostatics 1979 Inst. Phys. Confr. Series 48 1979 p85-95

[10] J N Chubb *"Measurement of tribo and corona charging features of materials for assessment of risks from static electricity"* Trans IEEE Ind Appl. 36 (6) Nov/Dec 2000 p1515-1522

[11] J N Chubb, P Holdstock M Dyer *"Predicting the maximum voltages expected on inhabited cleanroom garments in practical use"* 'Electrostatics 2003' 23-27 March, 2003. IoP Conference Series 178 2004 p131

[12] J. N. Chubb *"The assessment of materials by tribo and corona charging and charge decay measurement"* Inst Physics 'Electrostatics 1999' Confr, Cambridge, March 1999 IoP Conf Series 163 p 329-333.

[13] J N Chubb *"Experimental comparison of the electrostatic performance of materials with tribocharging and with corona charging"* at: http://www.infostatic.co.uk/cache/Tribo-corona-comparison.pdf

[14] J. N. Chubb, P. Lagos J. Lienlaf *"Electrostatic safety during the solvent extraction of copper"* J. Electrostatics 63 2005 p119-127

[15] J N Chubb *"Charge decay time prediction"* Paper to be published in J. Electrostatics in 2010

[16] J N Chubb *"Test method to assess the electrostatic suitability of materials for retained electrostatic charge"* Document prepared in 2004 for discussion as prospective British Standard. Available at: http://www.infostatic.co.uk/cache/JCITestMethod.pdf

[17] J N Chubb *"A standard proposed for assessing the electrostatic suitability of materials"* J. Electrostatics 65 2007 p607-610

[18] P Holdstock, M J D Dyer, J N Chubb *"Test procedures for predicting surface voltages on inhabited garments"*. EOS/ESD 25th Annual Symposium. Riviera Hotel, Las Vegas, Nevada, USA 21-25 September 2003. J. Electrostatics 62 2004 p231-239

[19] J N Chubb, P Holdstock M Dyer *"Predicting the maximum voltages expected on inhabited cleanroom garments in practical use"* 'Electrostatics 2003' 23-27 March, 2003. IoP Conference Series 178 2004 p131

[20] *"Methods for measurements in electrostatics"* British Standard BS 7506: Part 2: 1996

[21] IEC 61340-2-1 *"Measurement methods in electrostatics – Test method to measure the ability of materials and surfaces to dissipate static electric charge"* Feb 2000. International Electrotechnical Commission

[22] J N Chubb, G.J. Butterworth *"Charge transfer and current flow measurements in electrostatic discharges"* J. Electrostatics 13 1982 p209

[23] J N Chubb *"Measurement of charge transfer in electrostatic discharges"* J. Electrostatics 64 (5) 2006 p301-305

[24] H M Hyatt *"The resistive phase of an air discharge and the formation of fast risetime ESD pulses"* EOS/ESD Symposium, Dallas Texas 1992 p55

[25] J. N. Chubb; S. K. Erents; I. E. Pollard *"Radio detection of low energy electrostatic sparks"* Nature 245 No 5422 1973 p206

[26] G. J. Butterworth *"The detection and characterisation of electrostatic sparks by radio methods"* Electrostatics 1979 Inst. Phys. Confr. Series 48 p 97

[27] Bernier, J; Croft, G; Lowther, R *"ESD sources pinpointed by analysis of radio wave emissions"* Proc EOS/ESD Symposium 1997 Santa Clara, USA p83

[28] Lin, D; DeChiaro, L. F. Ming-Chung Jon *"A robust ESD event locator system with event characterisation"* Proc EOS/ESD Symposium 1997 Santa Clara, USA p88

[29] Lewis, B. von Elbe, G. *"Combustion, flames and explosion of gases"* Academic Press, New York 1961

[30] M Glor *"Electrostatic ignition hazards associated with flammable substances in the form og gases, vapours, mists and dusts"* 'Electrostatics 1999' Inst. Phys. Confr Series 48 p1999

[31] M Glor *"Ignition risks from static electricity – problem solved?"* 'Electrostatics 2007' Journal of Physics: Conference series 142 p19

[32] N. Gibson, F. C. Lloyd *"Incendivity of discharges from electrostatically charged plastics"* Brit J. Appl Phys. 16 1965 p1619

[33] A W Bright *"Electrostatic hazards in liquids and powders"* J. Electrostatics 4 1977 p131

[34] E L Dinham *"Criteria for electrostatic ignitions"* Inst. Phys. Meeting December 1977

[35] H Strawson, A R Lyle *"Safe charge densities for road and rail tank car filling"* 'Static Electrification 1975' Inst Phys Confr Series 27 p276-289

[36] J M Van der Weerd *"Can probe-to-cloud discharges be an ignition source for tanker explosions?"* Shell Report AMSR.0016.72 1972

[37] P Boschung, W Hilgner, G Luttgens, B Maurer, A Widmer *"An experimental contribution to the question of the existence of lightning-like discharges in dust clouds"* J. Electrostatics 3 1977 p303

[38] G. J. Butterworth, E. S. Paul, J. N. Chubb *"A study of the incendivity of electrical discharges between planar resistive electrodes"* 'Electrostatics 1983' IoP Confr Series 66 p185

[39] C Buhler, C Calle, S Clements, M Ritz, J Starnes *"Test methodology to evaluate the safety of materials using spark incendivity"* J. Electrostatics 64 2006 p744-751

[40] J N Chubb, P Holdstock *"Risks of ignition from Type D ungrounded FIBC"* J Electrostatcs 68 2010 p145-151

[41] ANSI/ESD STM 11.31-2001: Bags

[42] J N Chubb *"How effectively do materials shield against transient electric fields"* IEEE Industry Applications Sept/Oct 2002 p13-20

[43] J N Chubb *"Non-contact measurement of the electric field shielding performance and the resistivity of layer materials"* J. Electrostatics 64 2006 p685-689

[44] *"Protective clothing: electrostatic properties – Part 3: Test methods for measuring charge decay"* Draft prEN 1149-3 (3rd rev) Test method 2 (Induction charging)

[45] Smythe W R *"Static and dynamic electricity"* 1968 McGraw-Hill, New York

[46] Trowbridge, C. W. *"Computer modelling of electrostatic fields"* Electrostatics 1991, Inst. Phys. Confr. Series 118 p253

[47] Thomas, C. L. *"POTENT A package for the numerical solution of potential problems in general 2D regions"* Proc Confr on Software for Numerical Mathematics and its Applications, Loughborough Univ, April 1973 Academic Press, London

[48] Thomas, C.L. *"THREE-D A digital computer code for the design and analysis of three-dimensional electrostatic fields"* IEE Confr Computer Aided Design, Univ Southampton, April 1974

Chapter 5

ASSESSMENT OF SIGNIFICANCE OF STATIC

5.1. INTRODUCTION

There are many operations in industry where it is likely that static electricity will arise. This may cause risks or problems – or it may be the basis of constructive applications.

The two main questions which need to be considered are:

- What are the levels of parameters where static is likely to cause problems or be useful in particular situations?
- What measures can be taken to reduce the risk or severity of problems and/or ensure reliable application?

The first question, considered in this Chapter, concerns static levels (surface and volume charge densities, electric fields, surface voltages) at which static is likely to cause problems or be used. The second question, about remedial measures, is considered in Chapter 6. Having taken actions to control or use static there is then the need for measurements to confirm if the actions are working as required – and are continuing to work.

The main categories of problem with static electricity are:

- risks of ignition of flammable gases
- shocks to personnel
- attraction of dust and debris
- cling of thin films
- damage to semiconductor devices
- upset of operation of microelectronic systems

The main constructive applications of static electricity relate to electrostatic forces of attraction between opposite polarity charges – relevant to layers, thin films and particles. These are relevant, of course, in such major areas as photocopying and in air cleaning by electrostatic precipitation of dusts.

5.2. Ignition of Flammable Atmospheres

The risk from static electricity in flammable atmospheres is primarily related to the energy which can be released in a spark type discharge between conducting surfaces [1]. The minimum energies needed for ignition for a range of materials are:

- hydrogen/air 0.02 mJ
- common hydrocarbon vapour/air mixtures 0.2mJ
- powders 1mJ upwards

The electrostatic energy available from a charged capacitor with a direct spark discharge is:

$$U = \frac{1}{2} C V^2$$

In some situations, for example discharges from people, not all the available energy will be released in the discharge gap because energy is dissipated in resistance in the discharge path [2].

Incendive static discharges can also occur when a blunt earthed projection (radius of curvature around 12 mm) comes up to a charged insulating surface - for example a charged film [3]. If the charged surface is freely supported than it seems the surface voltage needs to be above 20-30kV and the quantity of charge transferred in the discharge more than 100nC [3]. Negative polarity surface charges are appreciably more incendive than positive [4]. If the charged film lies on an earthed backing then incendive discharges, which propagate over large areas of the film (propagating brush discharges), can occur if the surface voltage is greater than about +4kV [4].

A localised high voltage discharge will occur at a fine tip where the electric field locally exceeds the breakdown strength of air but the electric field over the remaining gap is too low (below around 200kV m^{-1}) to enable the discharge to propagate across the whole gap as a spark type discharge. Such 'corona' discharges are considered to be non-incendive, and offer the useful possibility of safely removing charge from charged conductors and insulators even in flammable atmospheres - and so acting to reduce risks.

If the discharge occurs at the end of a rather blunt projection (for instance around 12mm radius of curvature) then a 'brush' type discharge can occur. For this to be incendive it seems the potential of the projection needs to be over about 65kV [5]. While the energy dissipated in a spark type discharge between conducting electrodes can be calculated ($U = \frac{1}{2} C V^2$) it has so far not proved directly feasible to make comparable calculations to predict the incendivity of corona and brush discharges from measurement of electrical parameters.

Ignition risk mechanisms can be quite subtle. For example, the dipping of a silo or fuel storage tank with a conducting cord could create a risk if the cord and/or the operator are not earthed. Charge could be induced on the cord by space charge in the vessel with risk of discharge to the edge of the sampling aperture if the cord comes too close. In the operation of large crude oil tankers high pressure water jets are used to clean tank walls during the ballast voyage. The water mist is charged and because of the large size of the tanks space potentials up to 30-40kV can arise [6]. If tanks are not inerted there is a risk that slugs of washing water can acquire sufficient charge (and thereby electrostatic energy) from the high electric fields at tank structure edges that incendive sparks can occur when the slugs reach other parts of the tank structure [7] – see also Chapter 7.3.

Ignition risks can arise from the strong electrostatic charging which occurs when condensing gases are released through nozzles at high pressure - for instance in the release of steam or carbon dioxide [8,9]. Major ignition events have occurred with the use of carbon dioxide at attempted inerting operations where flammable gases were present (Alva Cape, Bitburg) [8].

5.3. Shock Risks

If a person becomes charged, for example by walking across a nylon carpet, a shock is likely to be felt on touching an earthed conductor if the body potential is above a few kV.

Discharge energy: (mJ)	Equivalent body voltage: (kV)	Response:
1	3.6	perceptible sensation
10	11.5	definite shock
100	36.5	unpleasant shock
1,000		severe shock, muscular contraction
10,000		possibly lethal shock

Maximum body voltages by walking on carpets, etc in dry atmospheres are in the range 20-30kV. Body voltages up to 20kV can occur at getting out of a car. Much higher voltages, and hence discharge energies, can arise electrostatically at, for example, reel up of charged webs (e.g. paper). Hovering helicopters can become highy charged and create shock risks to personnel lowered to the ground - as in a rescue situation. Shock risks can arise by direct discharge from the highly charged surface or indirectly by induction charging in the high field with the shock at subsequent earth contact.

Electrostatic shocks are unlikely to be lethal. They may be surprising, or unpleasant – but the main practical risk is from the consequence of the surprise reaction (e.g. falling off a ladder).

5.4. Attraction of Dust, Debris and Thin Films

When thin films with slow charge dissipation rates pass over rolls at speed or are peeled off a surface they are likely to become highly charged. There will then be a force of attraction to any nearby earthed surface (image force). These forces may cause mechanical problems in the handling of charged thin films. There will also be attraction of airborne dust and debris particles, which are usually charged to some extent. The collection of airborne particles on films or plastic mouldings may degrade the application of the product and/or spoil its cosmetic appearance. Appearance is very important in the retail trade. Degradation of suitability for application may also apply in such situations as bedding and nurse's aprons in hospitals.

A sheet of thin film with nett surface charge density σ (C m^{-2}) near an earthed plane surface experiences an attractive force F_e:

$$F_e = \sigma E = \sigma^2 / 2\, \varepsilon_o$$

The gravitational force on a thin film per unit area of density (ρ) and thickness (t), with g = 9.81N/kg, is:

$$F_g = \rho t\, g = 9.81\ \rho\, t$$

Cling will occur if electrostatic forces exceed gravitational forces. This occurs, for example, for a 20μ film when surface charge density exceeds about 2 $\mu C\ m^{-2}$ and the electric field at the nearby earthed surface is over about 300 $kV\ m^{-1}$.

5.5. Damage to Semiconductor Devices

The operational performance, and/or reliability, of semiconductor devices can be prejudiced if direct electrostatic discharges occur to the device itself or to connecting leads on circuit board tracks or if connections are exposed to transient high electric fields. Discharges may destroy on-chip connections or thin them down. The voltages created from transient high electric fields may breakdown thin chip insulating layers.

The susceptibility of devices varies greatly and techniques are improving to provide on-chip protection. On the other hand device geometries are getting finer and the 'real estate' needed for protection is non-trivial.

Risks are generally thought of in terms of a 'human body model'. The general feeling seems to be that in work areas where semiconductor devices and assemblies are handled electrostatic voltages should be less than 100V. For work with MR head components it seems that voltages may need to be down below 20V. A 'charged device model' is more appropriate when a device, which has become charged by sliding over a surface, or by induction in nearby electric fields, contacts earth or a high self-capacitance conducting surface. In this situation it seems the threshold voltage for damage may be around 25% of the corresponding 'human body model' value.

5.6. Upset Operation of Microelectronic Systems

Electrostatic discharges can involve current flows up to a few amperes and changes in local electric fields with frequency components over a very wide range - up to a GHz [10]. If such signals couple into the circuits of microelectronic systems it is not surprising that operations may be upset and/or data structures corrupted. A major part of the European EMC directive concerns achieving demonstrated immunity of equipment to signals over a wide frequency range. IEC 801.2 involves testing of equipment with a source of high voltage sparks that, of course, may penetrate case apertures and couple directly to circuitry. Equipment operation should be immune to discharges up to about 20kV - which is a sensible maximum value likely to be reached by people working in environments where static is not controlled.

As reliance on microelectronic systems becomes greater, as complexity increases and the range of applications widens, it will be increasingly important to avoid any risk of operational upset.

References

[1] Lewis, B. von Elbe, G. *"Combustion, flames and explosion of gases"* Academic Press, New York 1961

[2] N. Wilson *"The nature and incendiary behaviour of spark discharges from the body"* Inst Phys Confr Series No 48 1979 p73

[3] N. Gibson, F. C. Lloyd *"Incendivity of discharges from electrostatically charged plastics"* Brit J. Appl Phys 16 1965 p1619

[4] G. Luttgens, M. Glor *"Understanding and controlling static electricity"* Expert Verlag 1989

[5] J. M. van der Weerd Shell Report AMSR 0016 1972

[6] J. M. van der Weerd *"Electrostatic charge generation during washing of tanks with water sprays, II Measurements and interpretation"* Static Electrification Conference, London, 1971 IoP p 158

[7] J. N. Chubb *"Practical and computer assessments of ignition hazards during tanks washing and during wave action in part ballasted tanks"* J. Electrostatics 1 1975 p61

[8] E. Heidelberg, K. Nabert, G. Schon Arbeitsschutz 11 1958 p221

[9] G. J. Butterworth *"Electrostatic ignition hazards associated with the preventative release of fire extinguishing fluids"* Electrostatics 1979 Inst Phys Confr Series 48 p161

[10] H M Hyatt *"The resistive phase of an air discharge and the formation of fast risetime ESD pulses"* EOS/ESD Symposium, Dallas Texas 1992 p55

Chapter 6

PRACTICAL ELECTROSTATIC PROBLEMS

6.1. GENERAL COMMENTS

The preceding Chapters have provided a background of theory, methods for measurements, practical aspects of measurements and the levels at which static is likely to cause problems in different situations. It is now time to turn to practical questions:

- how do you find out if it is static causing the problem experienced?
- is static the real risk or problem in particular situations?
- what can be done to overcome/avoid risks?
- how do you know if remedial measures remain effective?

We start by noting the 5 Cs of static electricity:

C1: Characteristics Of Materials:

- chargeability
- charge decay
- capacitance loading
- resistivity
- dielectric constant
- shielding
- ability to support an incendive electrostatic discharge

C2: Coupling:

- geometry effects
- modelling
- time domain effects
- capacitance

C3: Consequences:

- electric fields
- surface voltages
- stored energy
- electrostatic forces (attraction for dust and debris and thin films)
- ignition capability (metal electrodes/dielectric surfaces)
- EOS/ESD damage risk to semiconductors
- EMC radiation/susceptibility (upset operation of electronic systems)

C4: Countermeasures:

- reduce charging
- promote charge removal
- design to overcome consequences

C5: Constructive Use:

- electrostatic forces (electrostatic precipitation, photocopying, paint and crop spraying, particle alignment e.g. flocking)

6.2. Is 'Static' the Cause?

The first point to consider in any practical situation are whether there are materials and activities that do or are likely to give rise to separated charges and what measurements can be made to examine the practical situation. In many cases visual inspection and experience will provide a very useful starting point. If plastic or artificial fibre materials are present, if materials are dispersed into the air, if sizeable conducting bodies (notably people) may not be electrically well linked to earth – then there could be problems. It may well be feasible to make helpful measurements on materials away from the practical situation but it is desirable, and often necessary, to make observations in the real situation. This gives the important opportunity for unexpected features to show themselves – not just what you think is relevant!

The above overview needs to lead to an assessment of the potentials, quantities of charge, forces and electrostatic energies available in relation to the practical situation and operations. From this assessment the significance of various electrostatic aspects can be judged and recommendations made on actions to overcome or avoid risks and problems. To this will be added recommendations on future actions to ensure that the remedial actions proposed are checked to ensure they are effective and remain so.

When considering problems with static electricity it is important to remember that electrostatics is only ever a part of the overall 'system'. Electrostatic aspects must be viewed in conjunction with many other practical factors. Thinking of the 'system' is important because not only does it help assessment of the role of static but it also provides the framework for considering possible alternative ways to tackle problems. Understanding the electrostatic aspects of problems is usually desirable, but may not be necessary for achieving a solution. The 'problem' may be an electrostatic problem: but the 'solution' need not depend on electrostatics. For instance: risks of ignition of flammable atmospheres by static discharges might be tackled by avoiding static. Alternatively, risks of ignition might be removed by avoiding the occurrence of a flammable atmosphere. In such situations it is important to recognise that there may be other mechanisms of ignition and to check for these - for example by frictional heating (trapped material rubbed in mechanical systems), impact sparks (for example steel and flint) or thermite reactions (involving aluminium or other light alloys with rusty steel).

It is important not to stop thinking of alternatives as soon as one possible cause or one solution of a problem has been identified. There may be several contributory factors and a variety of solutions. All factors need to be recognised and considered in the context of the 'system' - which will include personnel, engineering implementation and economics as well as technical aspects. There are, however, a number of general basic actions that need to be taken for reasons of safety [1,2,3].

6.3. Codes of Practice

A number of Standards relating to the control of static are listed in Chapter 7.6. The main Codes of Practice available that relate to static electricity concern:

- the control of static risks in relation to risks of shock, mechanical handling problems and ignition of flammable gases in petrochemical and processing industries [1,3]
- the control of static risks in the handling of semiconductor devices and microelectronic assemblies [2].

The basic philosophy for the control of static in the above Codes is to provide reliable earth linkage everywhere for everything. This to ensure that any static charges that arise will have opportunity to migrate quickly to earth and not be available for retention on isolated conductors or insulator surfaces. This means visible metal earth bonding straps for all metal plant. With plastics and naturally insulating materials it means avoiding charge retention by, for example, an alternative choice of material or by modifying the material by an additive in manufacture or by surface treatment to promote charge dissipation. Where charge retention cannot be adequately controlled by material selection then local air ionisation may provide a route for charge neutralisation by making the local atmosphere partially conductive.

It may be expected that static could be controlled by appropriate choice of interacting materials according to the 'triboelectric series'. This is not, in general, a practicable or reliable approach to avoid risks or problems – but it may help reduce them. It is easy to show that PTFE rubbed on PTFE gives two highly charged surfaces!

Codes of Practice cover many common practical risk situations very well, but there are a number of problems which are not covered. The following sections provide ideas of some alternative approaches worth consideration.

6.4. Characteristics of Materials

Resistivity measurements have traditionally been the way used to qualify materials where static electricity is thought likely to causes problems or presents risks. In some situations such measurements can be appropriate (e.g.

flooring and footwear) when the need is to drain charge from a conductor in contact (e.g. from the body to the floor). Where problems arise from static charge retained on a material itself then a measurement of resistivity may not be appropriate. This is particularly true for inhomogeneous materials. Resistivity indicates the fastest route available for charge migration. The slowest route for charge migration is what is relevant for charge retention. Charge decay measurement is appropriate in such situations. However, it is important that the method of measurement used is suitable and is shown to give results that match the decay of triboelectrically generated charge [4,5] (see also Chapter 3.4). It is to be noted that Federal Test Standard 101C Method 4046 does not achieve this – as noted in Chapter 3.4.3 [4]. To avoid the occurrence of local high voltages and to limit the time they may be present requires the charge decay time to be short compared to the time of mechanical actions responsible for charge separation. Decay times below ½ s have been suggested as suitable but recent studies have indicated that decay times below ¼ are preferable [5]. It is worth noting that decay time must not be too short (probably not less than a few ms), to avoid the possibility of spark discharges directly to the material itself.

Two particular areas where it is necessary that charge decay times are not short is in electrostatic precipitation of airborne particles from flue gases and in electrostatic paint spraying. If charge can migrate too quickly through the layer of particles built up on the precipitator collection electrodes (as applies with precipitation of, say, carbon black) then the material is not adequately held to the surface by electrostatic forces and will fall away prematurely and so become re-entrained within the airflow. If, on the other hand, charge cannot migrate adequately quickly through the layer of particles in relation to the local flow of corona current then a large voltage can build up across the layer. This will reduce the electric field depositing charged particles and hence particle collection efficiency It may also lead to electrical breakdown through the layer and cause 'orange peel' type non-uniformity in sprayed paint layers.

A special feature of materials that can limit the influence of retained static charge is the capacitance presented to charge on the surface. If the capacitance is high then the surface voltage likely to arise from the quantities of charge transferred in practical events may be sufficiently low as not to allow risks or problems. Measurement of this capacitance effect is discussed in Chapter 3.4.5 [6]. The following Figures show examples of how charge decay times and capacitance loading vary with humidity in different ways for a cotton fabric and for paper. Charge decay times for 'finished' papers seem fairly constant with humidity, whereas for newsprint and for a cotton fabric they are much longer with lower humidity. The constancy of decay times of 'finished' papers may arise because of a near balance between increasing charge mobility and increasing capacitance loading with humidity.

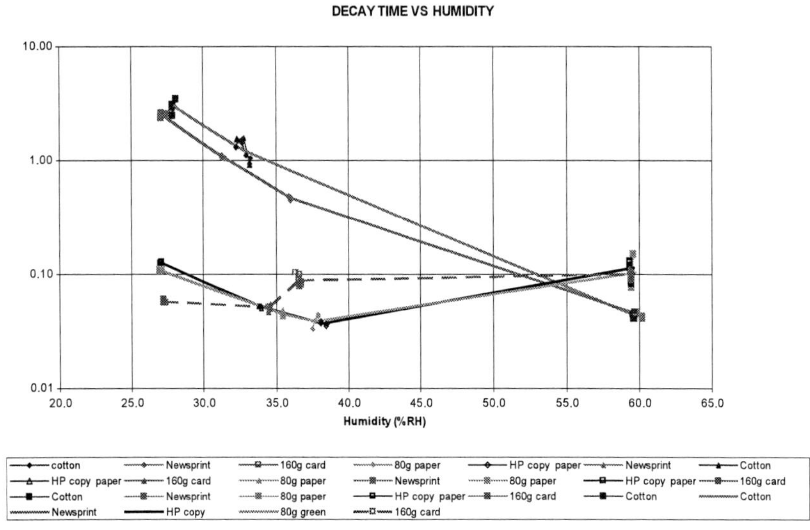

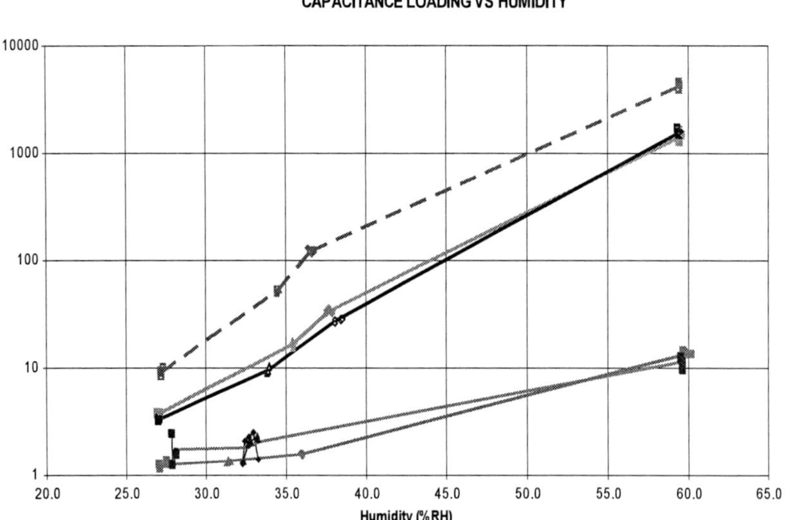

Figure 6.1: Variations of decay time and capacitance loading with humidity for a cotton fabric, newsprint and some finished papers and card

6.5. Investigative Procedures

The fieldmeter is the main instrument for investigation of electrostatic questions. These instruments can be used, as explained in Chapter 4, for measuring the main parameters of interest in electrostatic investigations. The following comments draw attention to a number of points to remember in studies in practical situations.

- *Sources of static:* A fieldmeter is useful to scan around a work area to identify materials and operations where static charge and surfaces voltages are present. Remember to earth bond the fieldmeter. For induction probe instruments (no rotating chopper) it is necessary to switch the instrument on with the sensing surface well shielded from any sources of charge. A high sensitivity fieldmeter (field mill type) makes it easy to find even low levels of static at a distance. When possible sources have been identified one can focus in to confirm identification and make quantitative measurement. It is necessary to be aware of influence of other surfaces nearby that may be charged or may be shielding the effects of charge. Where suspect materials are lying on another material, particularly on an earthed surface, it is helpful to lift the material to see whether it is this that is actually charged – and to try rubbing the material to see if it easily becomes charged. It is best not to scan too quickly when looking for sources of static as the response time of the instrument display may not show observations reliably.
- *Body voltage:* A fieldmeter in an electrostatic voltmeter configuration provides a useful basis for near zero current drain measurement of voltages. The electrostatic voltmeter can be connected to the person by a trailing lead to, for example, a wrist strap. The lead should have good quality insulation and not drag on other surfaces. Because body actions involve timescales to below a second it is wise to record observations with a time resolution better than ¼s .
- *Charge decay:* A simple approach to assess charge dissipation capability of powders is to slide the powder down a chute of the material likely to be relevant on to an earthed metal plate. A sensitive fieldmeter mounted above the pile of powder will show how strongly the powder is charged and how quickly this charge goes away. If the powder is very dusty then it is wise to shield the sensing aperture of

the fieldmeter while sliding the powder down the chute. The difficulty with the method is that it will not be very quantitative for the quantity of charge generated, or the size and form of the pile of powder, etc. This is where more specialist instrumentation (see Section 3.4) and a formal test procedure (see Annex 2) become appropriate.

- *Flooring:* Resistivity is the appropriate way to assess flooring. The overall ability of flooring and footware to control body voltage can be demonstrated by linking the person to a recording electrostatic voltmeter (see 'body voltage' above) and getting the person to walk, scuff their feet, get up from a chair, etc on the floor.
- *Garments:* The ability of garment fabrics to dissipate static charge and so avoid problems of cling, attraction of atmospheric dust and debris and risk of damage to microelectronic systems is most simply assessed by charge decay measurements (see Section 3.4). With fabrics for cleanroom garments the inclusion of conductive threads may provide sufficient capacitance to surface charge to enable the material to be suitable to avoid appreciable surface voltages even though the decay time may be long [6].
- *Microelectronics*: The problem in microelectronics is that the surface voltages and discharge energies permissible are very much lower than in any other area of electrostatics. This means that instrumentation for detecting and quantifying static around the work area needs to be that much more sensitive than for other areas of interest. Assessment of the suitability of materials equally needs to be that much more rigorous.

The question of proximity is also relevant in the selection of materials. If for practical convenience items arrive for assembly in 'unsuitable' packaging can one be sure to avoid this packaging coming close to sensitive devices and circuits? Note that devices do not become insensitive to static just because they are on a circuit board – you may have just connected the device to extensive aerials! Note also that solder resist coatings on PCBs can become charged and hold that charge! Where materials cannot be made 'acceptable' in themselves then other measures need to be considered – such as air ionisation as a useful additional way to control static.

6.6. SYSTEM CHANGES

The following notes suggest general ways risks and problems may be reduced or avoided by changes to the 'system':

a) Change system design to avoid any significant levels of static arising on isolated conductors - for example earth bonding all metal parts of plant, ensure operators wear conductive footwear and ensure floor free of insulating layers, avoid chance of release of flammable gases and dusts, use of inert gas where flammable materials can be released (but think about asphyxiation risks), change materials so static can leak away easily, on-chip protection in semiconductor devices to guard against static discharges, make microelectronic systems immune to influence of static discharges.
b) Design to accept consequences - for example explosion venting, explosion suppression, remove personnel from direct line of fire for loading reactors, avoid possible knock-on effects to prevent possible small events becoming major disasters (e.g. think through the 'What if...' possibilities)
c) Minimise static generation - avoid sliding of materials (webs, powders), avoid stalled rolls, limit speeds (processes, liquid flows), minimise use and areas of insulating material surfaces.
d) Enhance charge leakage - for example by choosing or modifying materials to have good charge decay characteristics. This may require use of humidity control or additives or surface treatments
e) Avoid the presence of the dust or debris that can be attracted to surfaces that may become charged – for example, by a local supply of well filtered air
f) Train staff to increase awareness of risks and benefits of following guidelines and requirements. (See general guidance from Health and Safety Executive).

6.7. Some Specific Examples

6.7.1. Risks At Loading Reactors Containing Flammable Solvents:

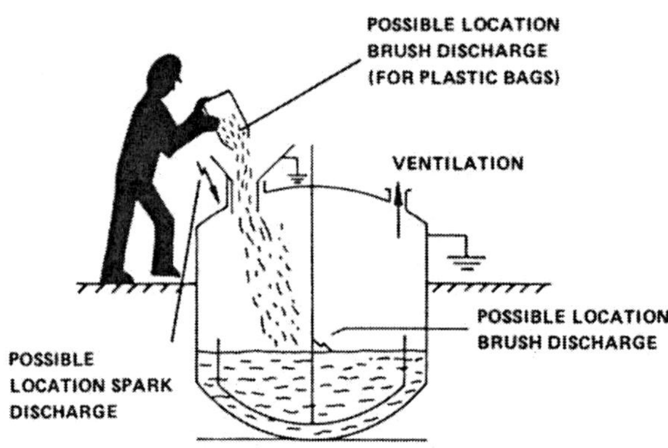

- opening hatch to over-rich reactor atmosphere may allow a region within flammable range around hatch
- stirring, or stopping of stirring, can cause generation of high voltages within reactor, on surface of liquid or on any unbonded conductors
- sliding of powders out of packaging, plastic or paper, is likely to create charged powder. This is likely to mean charged packaging and a charged operator if he is not adequately bonded to earth.
- incendive discharges can occur if highly charged packaging or a charged operator touches reactor hatch opening
- if ignition and explosion do occur the operator is in prime target position for damage!

Remedial approaches: Use of inert gas seems attractive but is not simple. Inerting may not be effective around open hatch and sizeable quantities of inert gas may be needed if frequent hatch openings occur. There is an asphyxiation risk to operators. Best to segregate operator from an open hatch using indirect loading via rotary loading valve and have a separate safe route blow off vent if any ignition does occur. If open hatch must still be used then best to use paper not plastic lined sacks, to provide general fresh air movement to clear flammable atmosphere around open hatch and to ensure operator is earthed via footwear and by contact to a support guard rail which he will naturally contact on loading into hatch opening.

6.7.2. In Silos and Flammable Liquid Storage Tanks:

All debris (in particular any cans that might float in fluid systems) need to be removed before filling vessels.

Direct manual lowering of a sampling cup into the vessel should be avoided. A conducting cup and/or line may collect charge and allow the occurrence of a spark on touching probing aperture. An insulated person may become charged via the support line and then discharge on touching the probing aperture where a flammable atmosphere may be present. Dipping and gauging should take place via a permanently mounted earthed shield tube and use only dissipative dipping or gauging equipment.

6.7.3 Transfer of Charged Material into Vessels/Containers

Care is needed when material is transferred into a vessel where the material is likely to be charged - as applies with many fine powders, The introduced charge will appear on the outside of the vessel if the vessel is not earthed. If high surface voltages build up on the surface of the transferred material this may upset transfer activity. It may also allow opportunity for direct shock to personnel whose hands approach the filled vessel.

The receiving container must be earth bonded. If the container has an insulating lining and/or the powder may have a long charge decay time then an earthing rod should be placed into powder within the lining. This will provide a resistive path for charge leakage and opportunity for corona charge leakage if surface potentials try to rise to a high level at the powder surface.

6.7.4. FIBCs

If charged powders are collected in 'fabric intermediate bulk containers' (FIBCs) the fabric should be dissipative and preferably have multiple conductive threads woven in. For Type C FIBC there is a requirement that this pattern of conductive threads shall be reliably bonded to earth. This is not an easy requirement with multiple repeat use of FIBC.

Type D FIBC use a pattern of resistive threads and ignition probe tests show that these can be made to avoid risks of ignition even when these FIBC are not bonded to earth. It is not yet fully clear how Type D FIBC materials avoid risks of ignition when they are highly charged and not bonded to earth. It is suggested that the inclusion of 'conductive' threads with quite high core resistivity may both limit the area of surface from which electrical discharges can draw charge and also limit the ability to draw charge from neighbouring areas [9].

6.7.5 Fluids

Charge is generated when high resistivity (long charge decay time) fluids flow through pipes or are stirred – or the fluid is otherwise subject to shear. This charge will be retained on fluids collected in vessels with insulating linings, so an earthing route must be provided by an earthing rod or plug of adequate area (as 6.7.3 above).

6.7.6. Moving Webs

Charge arises on moving webs after contact and separation from drive and handling rolls. This charge may be neutralised using passive, radioactive or active (corona) charge neutraliser bars. This only neutralises the charge locally. A better solution, where feasible, is to change the characteristics of the material to provide more rapid charge dissipation.

It needs to be noted that with webs moving at high speed (more than a few m s^{-1}) a 50/60Hz energized corona neutralizer bar is likely to leave bars of alternating charge on the web surface. This may be overcome by using a suitably high frequency energisation.

6.7.7 Flooring

Wherever flammable atmosphere may occur conductive/dissipative flooring and conductive footwear should be used to provide a route for dissipation of charges on personnel. In many industries flooring may get covered with powders that may be insulating – so other ways of earth bonding are needed (for example an earthed rail against which the operator is likely to lean during operations).

6.7.8 Humidity

It may be possible to use humidity to enhance electrical conductivity of film and powder products and surfaces. However, this may cause problems of adhesion, may be technically unacceptable for the product or may not be successful if product does not adsorb moisture.

6.7.9. Aircraft Refuelling:

The flow of insulating liquids (for example high quality hydrocarbon fuels) through fine filters can cause strong charging. Charge can accumulate on any unbonded conductors but also on free surface of fuel. It is usual to limit maximum flow rates and use antistatic additives where feasible [3]. The rate of charge dissipation in low conductivity fluids collected in large vessels is more important than initial charge density - charge decay time needs to be short compared to filling time.

6.7.10. Carbon Dioxide

The release of carbon dioxide causes strong charging when 'snow' is produced [5,6]. This can be a problem in preventative release of fire extinguishing gases [7,8]. Major accidents have been Bitburg and Alva Cape.

6.7.11 Aerosol Cans

Aerosol cans may become charged if leakage of the high pressure propellant occurs. If the propellant is flammable then there can be a risk of ignition. Lifting a leaking can from earthed surface can increase the risk by decreasing the capacitance and increasing the voltage - and hence increase the electrostatic energy available.

REFERENCES

[1] *"Code of Practice for Control of undesirable static electricity"* BS 5958: Part 1:1991

[2] *"Basic specification: Protection of electrostatic sensitive devices. Part 1: General requirements"* EN 100015: 1992

[3] *"Static electricity: Technical and safety aspects"* Shell Safety Committee 1988

[4] J N Chubb *"Comments of methods for charge decay measurement"* J. Electrostatics 62 2004 p73-80

[5] J N Chubb *"Experimental comparison of the electrostatic performance of materials with tribocharging and with corona charging"* http://www.infostatic.co.uk/cache/Tribo-corona-comparison. pdf

[6] J N Chubb *"A standard proposed for assessing the electrostatic suitability of materials"* J Electrostatics 65 2007 p607-610

[7] E. Heidelberg, K. Nabert, G. Schon. Arbeitsschutz 11 1958 p 221

[8] G. J. Butterworth *"Electrostatic ignition hazards associated with the preventative release of fire extinguishing fluids"* Electrostatics 1979 Inst Phys Confr Series 48 1979 p161

[9] J N Chubb, P Holdstock *"Risks of ignition from Type D ungrounded FIBC"* J Electrostatics 68 2010 p145-151

Chapter 7

EXAMPLES FROM EXPERIENCE

7.1. INTRODUCTION

A number of practical electrostatic studies are outlined in this chapter. These illustrate a range of approaches and observational techniques that can be involved together and the importance of making continuous recordings of the parameters observed.

7.2. CLEANROOM CLOTHING

The clothing worn in cleanrooms is basically a polyester fabric. This is needed for avoiding the shedding fine fibres and as an effective barrier for fine particles from the body and underclothing. Much clothing used for personal protection is based on the use of artificial fibres. These materials are likely to easily become electrostatically charged when rubbed and to retain that charge for an appreciable time. Antistat surface treatment is not acceptable in the case of cleanroom garments and would involve retreatment after laundering. Conductive threads have been included in the fabrics for these garments to provide a way to control static. Some conductive fibres are surface conductive, others are core conductive (with the conductive component within a polyester sheath). Fabrics with core conductive threads are preferred in practice because they retain performance over longer periods and there is less chance of shedding particles of the conductive component. The question that arises is how to assess the suitability of these materials and these garments for avoiding risks from static electricity? Traditionally the suitability of materials has been assessed by measurement of surface resistivity. This would reject fabrics with core conductive threads. With fabrics including surface conductive threads it is clear that if these threads contact the resistivity measuring electrodes that the measurements will refer to the resistance of the threads and will not provide any information on the areas of fabric between the threads. Recent studies [1] have shown that the initial peak voltage, per unit of charge transferred, does not depend on whether surface or core conductive threads are used – only on the pattern and spacing of the threads.

Corona charge decay measurements (see Chapters 3.4 and 4.9) are promoted as an appropriate way to provide a fair assessment of the ability of the fabric surfaces to dissipate static charge. Simply on the basis of charge decay measurement it is likely that any polyester fabric, even including conductive threads, will be considered unsuitable because charge decay times are likely to be long. However, if the problem is approached from the practical use point of view one asks the question as to what surface voltages are likely to arise from contact and rubbing actions. The assessment needed is then in terms of whether the surface voltages that arise will be significant. Thinking of assessment from this point of view made it clear that if charge transferred to a surface experienced a high capacitance then only low surface voltages would arise. The core conductive threads might provide an adequately high capacitance – and thus render the fabric and garments acceptable.

A number of studies were carried out that aimed to model practical use of garments [1,2,3,4]. An operator clothed in the garment to be tested stood on a plate connected to a virtual earth charge measurement unit. A fieldmeter was mounted a set distance (100mm) from the upper arm. Testing involved a single quick tribocharging scuffing action on the area of the garment directly under the fieldmeter with either a charge neutral Teflon rod or a woolen fabric mounted on an earthed wooden spoon. The person doing the tribocharging was clothed in a static free garment and bonded to earth. Observations were recorded of the quantity of charge transferred and the local surface voltage created and how this voltage decayed with time. From these studies the surface voltage per unit of charge transferred was calculated as well as the charge decay times. Measurements were also made of the corona charge decay characteristics of these fabrics at the same time under the same environmental conditions using a JCI 155v5 Charge Decay Test Unit. These studies included measurements of the initial peak voltage achieved per unit of corona charge transferred.

An example of comparison between tribocharging and corona charging performance is shown in Figure 7.1. It will be noted that the peak of surface voltage with tribocharging occurs about 100ms after the end of charging – shown by the start of the sharp negative swing as the rubbing surfaces separate and the fieldmeter reading is dominated by the influence of charge on the rubbing material before this moves quickly out of the way. Fair comparison with corona charging to the characteristics observed with tribocharging hence requires use of the surface voltage achieved a similar time after the end of corona charging – for example 100ms. With this approach the performance exhibited with tribocharging of garments matches fairly well to that exhibited with corona charging [5]. This approach to the assessment of materials was included in a document drafted for British Standards committee GEL101 [6] (see also Annex 2) and is the subject of a published paper [7].

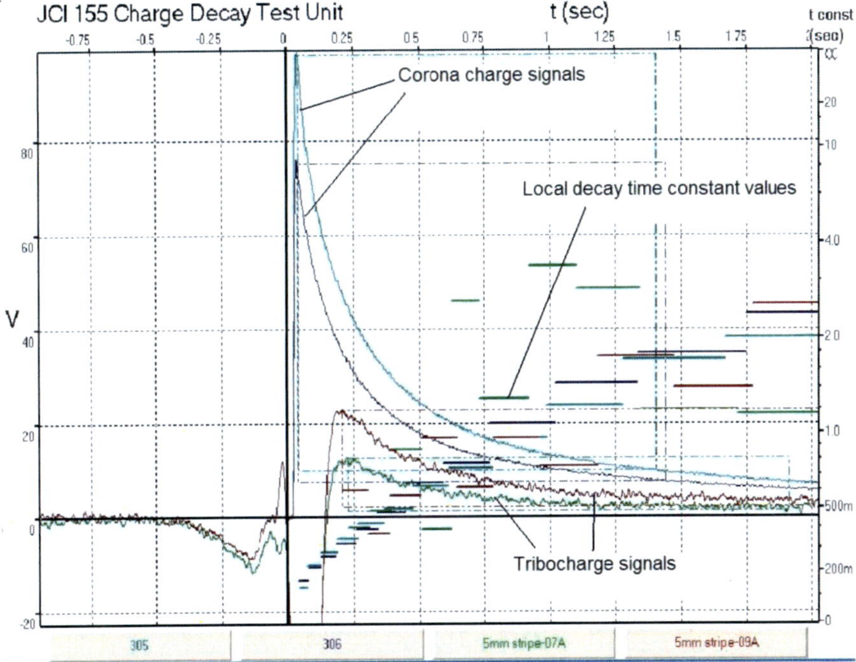

Figure 7.1. Response to corona and tribo charging of cleanroom garment fabric with 5mm stripe pattern conductive threads .

The outcome of the work with inhabited cleanroom garments was the concept that the capacitance experienced by charge on the surface of a material is a relevant feature for assessing the suitability of materials for avoiding risks and problems [2,3]. These studies also showed that it was possible to predict the maximum surface voltage that would be expected from tribocharging in practice from the results of corona charging measurements on the fabric.

A number of studies were also made where stretched and freely supported samples of various materials were tribocharged by impact with a Teflon rod in a similar way as with the inhabited garments [2,8]. The arrangement for this is shown in Figure 7.2.

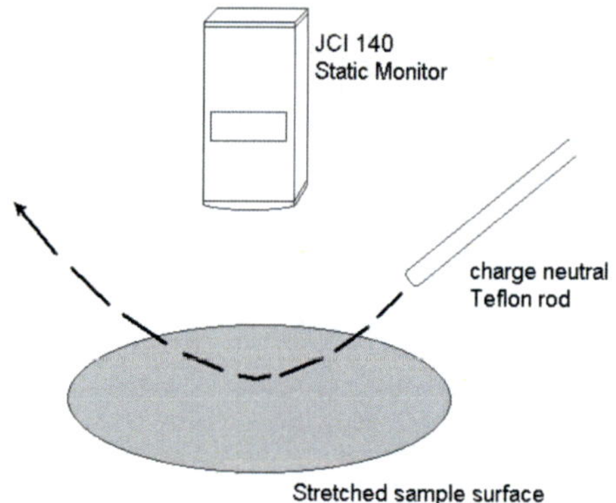

Figure 7.2. Arrangement for impact charging studies.

Examples of the relationships between the quantity of charge transferred and the initial peak voltage are shown in Figure 7.3.

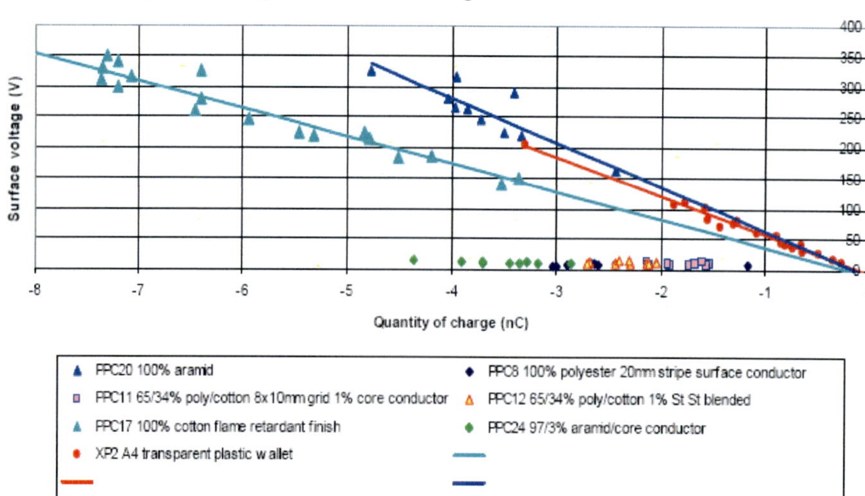

Figure 7.3. Variation of initial surface voltage with quantity of charge transferred.

Figure 7.3 shows very clearly in that the initial surface voltage varies linearly with the quantity of charge transferred and that the initial surface voltages can be strongly suppressed by the presence of conductive threads in fabrics. More recent studies have been based on tribocharging by dropping a ball on to an inclined stretched fabric surface and receiving the ball into a Faraday Pail for direct charge measurement [1].

7.3. Tanker Studies

In 1967 and 1968 three of the then new very large crude oil carriers (VLCC) suffered major damage from explosions. An investigation was started to find the cause and an appropriate solution. The explosions had all occurred during the ballast return voyage while the cargo tanks were being cleaned with high pressure water jets. Static electricity was suspected as a possible cause. An international research effort was mounted and many of the major oil companies were involved as well as a number of research institutes and universities. As the author, at UKAEA Culham Laboratory at the time, had some experience with static electricity (from PhD work) it was suggested that a contribution might be made as part of the laboratory diversification programme at that time.

Several of the other groups concerned with this problem were pursuing laboratory studies and some others, notably Shell KSLA in Amsterdam [9], were making some shipboard studies. Relevant laboratory facilities were not available at Culham and it was felt that what was important was to try to find out what actually happened in practice on board tankers during tank washing operations. The idea was put forward (perhaps not uniquely) that the mechanism for causing ignition of flammable gases in the cargo tanks was that slugs (lumps) of water created by action of the washing jets were charged by the electric fields formed from mist in the tanks and then being discharged with spark discharges as they reached other parts of the tank structure. It seemed that the occurrence of such spark events should be able to be monitored by radio observations. This proved to be a fruitful line of investigation and laboratory studies soon showed that radio observations at 38MHz were a very sensitive way to monitor the occurrence of spark events – with very low sensitivity to corona [10,11]. It seemed sensible that to avoid risks of making false observation by events local to a single radio aerial that two aerials should be mounted within the electrically shielded test environment and monitored via coincidence circuits with anti-coincidence to any radio signals observed at the same time by external aerials. Test equipment was built and its operation demonstrated in a large test tank at KSLA in Amsterdam and then during a pre-commissioning voyage of a new OBO (oil or bulk ore carrier). This OBO had been built for John Houlder (Houlder Bros) and who was motivated to contribute to the investigation of tanker explosions – not least because problems had also been experienced during ballast sloshing in the full width tanks of OBOs.

The above work lead to several shipboard investigations during ballast voyages [12,13]. The equipment was refined to include fieldmeters suitable for monitoring and interpreting the electrostatic conditions in the tanks during washing [14] (see also Chapter 4.16 on Modelling), radio detection observations on the occurrence of sparks and flash photography to try to photograph the situations in the tanks at the instant of occurrence of individual spark events (as illustrated in Figure 7.4 and 7.5 below). All the equipment had to be suitable for use in what might be a flammable atmosphere and also able to withstand the impact of high pressure jets of salt water!

The overall outcome of the shipboard studies was that sparks do occur during tanks washing operations, that there are large numbers of sparks and that they are associated with particular directions of impact of the washing jets around the tank structure. These shipboard studies were backed up by a number of laboratory investigations into the radio emission characteristics of sparks from various isolated bodies, by studies on the incendivity of discharges from bodies approaching a water surface and by a number of computer modeling studies [12,13].

Figure 7.4. Example of flash photograph during washing triggered by occurrence of a spark

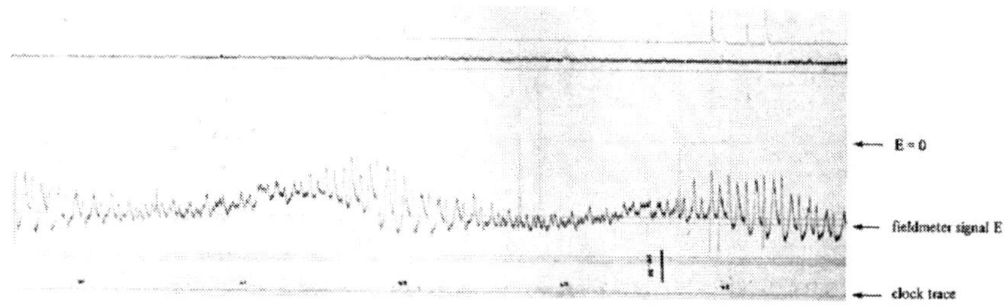

Figure 7.5. Example recording of fieldmeter signal variations during tank washing with associated occurrence of coincident radio signals from spark events (top trace)

It had been hoped that an understanding of how sparks occurred during tank washing might lead to mechanical design changes to minimize or avoid risks. However, the practical solution adopted by the industry has been to use inert gas (spent boiler gases) that is back filled during cargo discharge, to prevent the occurrence of flammable atmospheres within the cargo tanks. Conveniently, the volume of flue gas created during cargo discharge is adequate to achieve this back filling operation.

7.4. Food Product Silo

Anxiety had been expressed in 1978 about electrostatic safety in the operation of large food product silos following a number of explosions in grain elevators [15] that caused serious damage. A food product company operating a large silo in the UK felt it desirable to have a safety assessment made and this was taken on by the author.

It was appropriate for this assessment to use the instrumentation and approaches developed for the studies of tank washing of large crude oil tankers. From the top surface of the silo a couple of fieldmeters, on 50 m, connection cables were used to monitor the local potential near the top of the silo volume and to examine the potential distribution within the silo volume during and after filling activity. A couple of radio detection aerials were positioned near the top of the volume of the silo to pick up any occurrence of spark type discharges during silo operation. As in the tanker studies above (section 7.3) all observations were continuously recorded.

Examples of potential probe observations in the volume of the silo are shown in Figure 7.6 below during product filling and afterwards [13]. Figure

7.7 shows an example of sudden voltage excursions that were observed on a few occasions during these studies. These were observed with both fieldmeters so were clearly real events – although no reason for their occurrence was established.

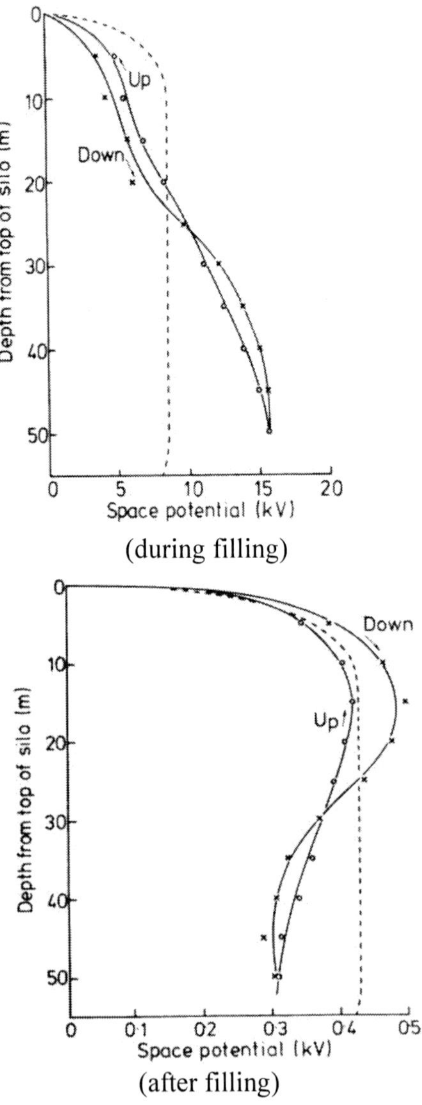

Figure 7.6. Potential distributions by lowering fieldmeter though volume of food product silo

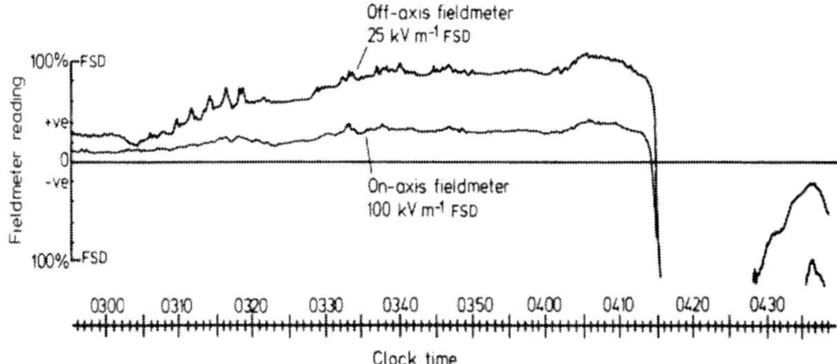

Figure 7.7. Fieldmeter observations made near top of food product silo during 'normal' filling showing occasional very large sudden excursions of space potential (100kV m^{-1} is about equal to 11kV of local space potential).

The outcome of these studies was that while there were these several quite large voltage excursions (tens of kilovolts) there were no radio signals observed to indicate the occurrence of any spark discharges. Although the sudden voltage excursions were electrostatically interesting it was concluded that there was no reason to be anxious about electrostatic safety. This would be true so long as there were no electrically isolated conductors present within the silo volume – such, for example, as drinks cans.

7.5. Lightning Warning

The local risk of occurrence of lightning depends on the local atmospheric electric field. However as this can change quite quickly a practical warning system needs to combine measurement of electric field with measurements of radio noise and lightning impulse activity to provide useful advance warning of risk – as discussed in Chapter 4.12 [16]. Practical systems need to be able to operate continuously in the presence of heavy rain, so instrumentation has to be designed and constructed appropriately and to incorporate operational health monitoring to ensure confidence in observations – as discussed in Chapter 3.2.7.6. Such a system was installed on the island of St Kilda (about 40 miles into the Atlantic off the Outer Hebrides) and has operated there for over 10 years. A second, more up to date system, was installed at Benbecula, in the outer Henrides, in the summer of 2008.

7.6 Car Seats

Many people experience shocks when getting out of their car. This is a nuisance and can be upsetting. The energy involved when the person's body discharges to the car bodywork can be sufficient to ignite flammable hydrocarbon vapour air mixtures. Such discharges have been responsible for a number of ignitions and significant damage in situations where vehicles have been refuelled with latching fuel fillers on epoxy painted forecourts. The problem here is that the customer may leave the filler in operation, get back into the vehicle, become charged on leaving the vehicle and create a spark discharge at contact with the filler in the flammable atmosphere issuing from the tank.

Measurements of body voltage were made on getting out of a car with various types of clothing and a number of different seat coverings on to a layer of insulation. An electrostatic voltmeter (JCI 148) was rested on the passenger seat and connected to a digital recording oscilloscope (Picoscope and laptop PC). The high voltage lead was connected to the wrist and arranged so that the person could easily get out of the car while still reliably linked to the electrostatic voltmeter. As shown in Figure 7.8 the body voltage rises to a peak and then falls back to a plateau level – the fall being associated with the second foot coming down to rest on the ground and the consequent increase in capacitance. These studies showed that even with normal cotton or wool based clothing and ambient environments around 50%RH it was easy to get body voltages in the 10-15kV range. This is enough to give quite a sharp shock.

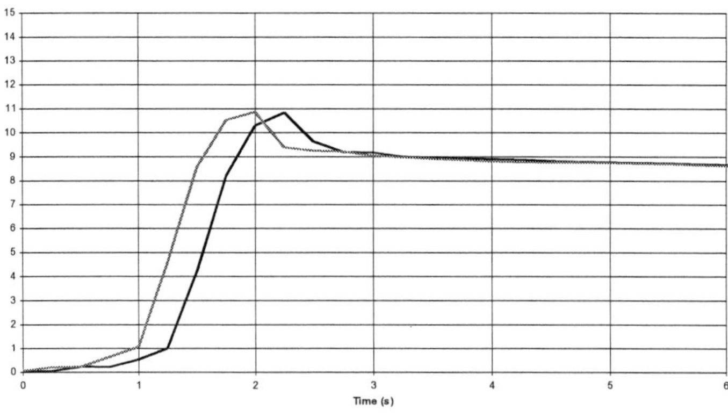

Figure 7.8. Variation of body voltage getting out of a car – normal car seat, wool suit (09/03/97).

The outcome of these studies was first, that body voltage measurements are reasonably reproducible test to test; and second, that there seemed prospects that maximum body voltages could be limited towards a level at which no shock would be felt (and at which there would be no risk of ignition) by using a seat covering that included a pattern of core conductive threads. Further progress depends upon the readiness of vehicle manufacturers to admit there is a problem - and that it is worth paying just a little for a solution!

7.7. Electrostatic Safety At Solvent Extraction Plant

There have been fires at a number of organic solvent extraction plants associated with the extraction and purification of metals. Such fires have caused extensive damage to processing plant that required serious expense for rebuilding. Static electricity has been suspected as a possible cause of ignition.

On-site studies have been carried out on electrostatic safety at a number of solvent extraction plants around the world. The aim was to investigate risk prospects and recommend remedial measures. The studies involved conductivity and charge decay measurements on process liquids sampled from various points around the plant, measurements of surface voltages on pipework and flanges and measurements of body voltages on walkways, etc [15].

Although organic solvent process liquids have high flashpoints any foam on a liquid surface can have quite a low minimum ignition energy. Foam is present on liquid surfaces in separation tanks, and is likely in part filled pipework so there is need to avoid any risk of electrostatic discharges.

The conductivity of the process liquids plays a major role in keeping voltages low throughout process plant - and so ensures safety in normal plant operation. Conductivity needs to be greater than 300pS m^{-1} (or charge decay times less than 0.05s) for flow velocities up to 2m s^{-1}. At higher flow velocities higher conductivities are needed. As in many situations involving the flow of hydrocarbon liquids, maximum flow velocities should be kept down to a few meters per second.

On HDPE pipework (used to avoid corrosion problems) only low voltages were observed. This would be expected so long as pipes are full of process liquid because of the capacitance to the liquid suppresses surface voltages. Such pipework hence does not present a significant risk. Metal flanges following pumps, and after any flow restriction, need to be earth bonded because if these were at high voltages there would be opportunity for metal to metal sparks to occur. Discharges to HDPE pipes or to process liquids themselves are not likely to create an ignition risk. Sampling of liquids on open surfaces in settling tanks needs to use insulating cups on insulating cord.

The assessment of plant safety requires proper measurement of the conductivity (or charge decay) performance of process liquids from various points around the operating plant. It is appropriate to make measurements on feedstock liquids in case there could be a flush though of these at any time. In addition, it is necessary to measure local voltages from time to time on pipework and any isolated metal flanges and valves around the plant. An appreciation also needs to be made of the consequences of fault or failure situations, indirect possible sources of risk and how these will be handled. All of this needs to be documented and audited on a regular basis.

7.8. Airborne Particles

The influence of electric fields on fine particles suspended in the air can conveniently be studied by photography using dark ground illumination with a defined width of illumination thrown at right angles to the direction of camera view. An exposure of say 1/30s will show the concentration of particles via the area concentration on the recorded images and the depth of the field of illumination. The direction and velocity of movement is shown by the track length and the exposure duration. Particles down to 5 microns can be tracked in this way.

The above approach was used in PhD thesis work [18,19] studying the relative influence of electric field and ionic wind on the behaviour of airborne particles in corona discharge fields. It was also used in an industrial study of the movement of airborne particles liberated at slitting of asbestos based engine gasket material and in a monitor for airborne fibres [20]. More sophisticated studies have been reported recently based on use of Doppler laser observations [21].

7.9. Airborne Fibre Monitor

A novel instrument was developed to enable the concentration and length and diameter information to be obtained on airborne fibres. This was aimed as the basis for a portable instrument for on-site assessment of the risks that might be presented by airborne asbestos fibres in workplace environments.

The approach devised [20] was to use a high frequency electric field to align of any asymmetrical airborne particles with laser light scattering arranged to selectively detect the presence of individual aligned particles in a streamline airstream within a clean air sheath. A high frequency electric field was used to avoid lateral movement of particles carrying a net charge. If a particle was detected by the selective scattering of light into one of two photomultiplier detectors the direction of the aligning field was switched through an angle of around 80 degrees. If the particle detected was a fibre then the signal at the first photomultiplier will decrease and then after a delay a signal will be observed at the second photomultiplier. The time taken for the second signal to rise relates to the length diameter ratio of the fibre. Experimental studies showed that fibres could be detected well down into the respirable size range with good immunity to other types of particle [19]. While prospects looked good technically, it was clear that a lot of testing would be needed to displace microscope analysis of personal monitor samplers. With modern image recording devices and signal processing capability there should be opportunity to identify the shape of a variety of asymmetrical airborne particles.

7.10. Coated Metal Sheets

Problems were reported with the handling of stacks of thin steel sheets that had a protective plastic film coating on one side. It was thought that static electricity might be binding the sheets together. Measurements showed that surface voltages on the coated films arising from sheet handling were fairly modest – only in the range 100-200V. These measurements also showed that these voltages only decayed away very slowly. Calculations soon showed that the area density of charge was actually quite high because it was the high capacitance between the outer surface of the coating film and the underlying metal surface that kept the surface voltage low. Proximity of the surface layer of charge to the smooth reverse surface of the next sheet above could easily generate forces of several kilograms – quite adequate to support the weight of a sheet!

Plausible solutions to such problems are a) to ensure that charge on the surface of the coating can dissipate, at least laterally, in a timescale short compared to plate handling times, and b) provide some topographic roughness of the coating to decrease the area of the film charged by rubbing and in close proximity to the neighbouring smooth metal surface in contact.

7.11. Adhesive Manufacture

The manufacture of adhesives involves the blending of solvents with a number of materials - some of which may generate very high levels of charge during mechanical preparation and mixing. Two approaches were developed to avoid the risks of ignition in such operations [22]. First, changing the sequence of mixing so that flammable atmospheres were not present when there were high levels of static charging. Second, use of antistatic additives (where permissible) to promote charge dissipation.

7.12. Health Monitor

Operation of a 24 hour body portable health monitor unit was found to be easily upset by static discharges occurring with normal daytime activities, such as removal of artificial fibre clothing at night. This was a particular problem in low humidity environments. The unit was basically a small belt mounted box of electronics to which a number of sensors over the body were connected. Signals from nearby static discharges would be picked up on the sensor leads, acting as aerials, and thereby coupled into the circuits and o available to upset circuit operation. Opportunities were available within the box for coupling these interference signals around the circuit by resistive, capacitive and magnetic induction mechanisms. The immunity to static discharges was raised to a very acceptable level by the simple expedient of capacitively decoupling each incoming lead directly to the metallised 'clam-shell' case of the box. This diverted the fast transient flows of charge to the lumped self-capacitance of the case and thereby avoided penetration of transient signals to the inside of the box. As more reliance is placed on microelectronic and computer control of industrial processes there needs to be greater appreciation of the risks of operational upset by static discharges and of approaches to deal with them.

REFERENCES

[1] J N Chubb *"Tribocharging studies on inhabited cleanroom garments"* J. Electrostatics 66 2008 p531-537

[2] J N Chubb, P Holdstock M Dyer *"Predicting the maximum voltages expected on inhabited cleanroom garments in practical use"* 'Electrostatics 2003' 23-27 March, 2003. IoP Conference Series 178 2004 p131

[3] J. N. Chubb, P. Holdstock and M. Dyer *"Predicting the maximum surface voltages expected on inhabited cleanroom garments in practical use"* Paper presented at ESTECH 2003, Contamination Control Division, Phoenix, Arizona.18-21 May, 2003

[4] P Holdstock, M J D Dyer, J N Chubb *"Test procedures for predicting surface voltages on inhabited garments"*. EOS/ESD 25th Annual Symposium. Riviera Hotel, Las Vegas, Nevada, USA 21-25 September 2003. J. Electrostatics 62 2004 p231-239

[5] J N Chubb *"Experimental comparison of the electrostatic performance of materials with tribocharging and with corona charging"* at: http://www.infostatic.co.uk/cache/Tribo-corona-comparison.pdf

[6] J N Chubb *"Test method to assess the electrostatic suitability of materials for retained electrostatic charge"* Document prepared in 2004 for discussion as prospective British Standard. Available at: http://www.infostatic.co.uk/cache/JCITestMethod.pdf

[7] J N Chubb *"A Standard proposed for assessing the electrostatic suitability of materials"* J. Electrostatics 65 2007 p607-610

[8] J N Chubb *"New approaches for electrostatic testing of materials"* J Electrostatics 54 (3/4) March, 2002 p2

[9] J. M. Van der Weerd *"Electrostatic charge generation during washing of tanks with water sprays, II Measurements and interpretation"* Static Electrification Conference, London, 1971 IoP p 158

[10] J. N. Chubb; S. K. Erents; I. E. Pollard *"Radio detection of low energy electrostatic sparks"* Nature 245 No 5422 1973 p206

[11] G. J. Butterworth *"The detection and characterisation of electrostatic sparks by radio methods"* Electrostatics 1979 Inst Phys Confr Series 48 p 97

[12] J. N. Chubb *"Practical and computer assessments of ignition hazards during tank washing and during wave action in part ballasted OBO cargo tanks"* J. Electrostatics 1 1975 p61

[13] J. N. Chubb, G. J. Butterworth, *"Instrumentation and techniques for monitoring and assessing electrostatic ignition hazards"* Electrostatics 1979 Inst Phys Confr Series No 48 1979 p 85

[14] I. E. Pollard; J. N. Chubb *"An instrument to measure electric fields under adverseconditions"* Static Electrification Conference, London, 1975. Inst Phys Confr Series 27 p182

[15] J N Chubb *"Methods for examining electrostatic ignition hazards"* International Symposium on Grain Elevator Explosions, National Academy of Sciences, Washington DC. July 11-12, 1978

[16] J. N. Chubb; J. Harbour *"A system for the advance warning of lightning"* Proceedings of Electrostatics Society of America Annual Meeting 2000, Brook University, Niagara Falls, Ontario, Canada. June 18-21 2000

[17] J. N. Chubb, P. Lagos, J. Lienlaf *"Electrostatic safety during the solvent extraction of copper"*. J. Electrostatics 63 2005 p119-127

[18] J N Chubb *"Behaviour of airborne particles during processes relating to dust collection"* PhD Thesis, Univ Birmingham 1958.

[19] J N Chubb, J. B. Higham and W. D. Bamford *"Experimental studies of airborne particle behaviour in corona discharge fields"* IEE Coll on Electrostatic Precipitators, Feb 1965

[20] J N Chubb *"A novel instrument to monitor and size classify airborne fibres in the respirable size range"* Electrostatics '91 IoP Confr Series 118 p169

[21]] M K Mazumder *"Measurement of particles size and electrostatic charge distribution on toners using E-SPART Analyser"* IEEE Trans Ind Appl 27 (4) 1991 p611-619

[22] J. N. Chubb, J. Embury, D. Bourne, S. Southall *"Measurements for the assessment of ignition risks from static electricity"* IEE Colloquium on Electrostatic problems during materials handling, London, Feb 15, 1994

Chapter 8

SOURCES OF INFORMATION ABOUT STATIC ELECTRICITY

8.1. GENERAL

Internet Websites provide a wealth of information about static electricity and about instrumentation and product suppliers. One may search under electrostatic, static electricity, fieldmeters, charge decay, capacitance loading, etc. and for individual companies and organisations (Google is a useful 'search engine'). Many companies and organisations involved with static electricity have their own Websites, and many include links to other relevant sites. However, many of the websites for companies offering instruments for electrostatic measurements provide rather little support or technical background information.

8.2 ORGANISATIONS ORGANISING CONFERENCES AND MEETINGS

Electrostatics Group, The Institute of Physics, 76 Portland Place, London, W1N 4AA (Tel: +44 (0)171 470 4800 Fax: +44 (0)171 470 4848 email: conferences@iop.org Website: http://www.iop.org/IOP/Groups/SE/). Concerned with basic research, atmospheric electricity, biological aspects, measurements, hazards and applications. 2-3 Group meetings a year and an International Conference in UK every 4 years: 1971 - 2007. Conference proceedings are published by Institute of Physics (Contact: IoP Publishing, Dirac House, Temple Back, Bristol, BS1 6BE)

A European Conference on Static Electricity is held every 4 years (alternates bi-annually with Institute of Physics Conference). These are held in different countries throughout Europe - in 2001 it was in Koscielisko, Poland. Selected papers are published in special issues of *Journal of Electrostatics.*

Institution of Electrical Engineers, Savoy Place London, WC2R 0BL (Tel: 0171 240 1871) organises occasional meetings and colloquia.

EOS/ESD Association (7900 Turin Road, Bldg 3, Suite 2, Rome, NY 13440-2069,USA Fax: +1 315 339 6793 website: www.esda.org) holds a Symposium each autumn in US - mainly concerned with interaction of static with microelectronics. Conference papers are published. ESDA is also responsible for generation of formal Standards in the US.

IEEE-IAS (Institution of Electrical and Electronic Engineers - Industrial Applications Society) - organises a conference in US each year, early October, with a section on electrostatics. Concerned with measurements, hazards and applications. Selected papers are refereed and published.

Electrostatics Society of America (ESA). Organises a Conference each summer. Their Website is at: http://www.electrostatics.org).

Meetings of other organisations which may have relevance are: The Dielectrics Society, International Aerospace and Ground Conferences on 'Lightning and Static Electricity'.

8.3. Information on Meetings

A list of meetings likely to be relevant to people interested in electrostatics is kept in the Journal of Electrostatics (http://ees.elsevier.com/elstat)

8.4. Journals for Publication of Papers

Journal of Electrostatics - published 6 times a year by Elsevier. Editor Mark Horenstein (email: mnh@bu.edu). The contents of the Journal for the last several years are listed on the Journal Website: http://ees.elsevier.com/elstat/

There is an on-line 'ESD Journal' at: http://www.esdjournal.com/

The EOS/ESD Association publishes a Newsletter 'Threshold' every other month on their Website.

8.5. DIRECTORIES

Kompass, Kelly's, Dial Industry
Internet: Google, Applegate

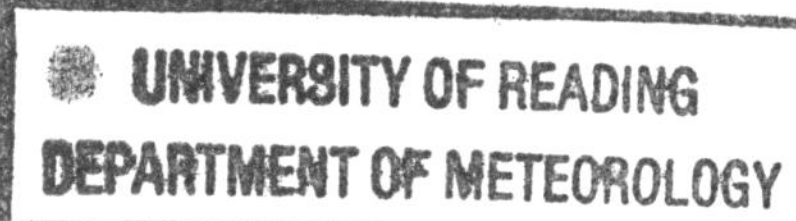

8.6. STANDARDS

There are a number of national and international organisations responsible for and involved in the preparation of formal Standards. Some 'Standard' documents are well presented and useful – others do not reflect best present appreciations of methods of measurement and/or assessment. A number of Standards continue to be referenced while out of date and/or outside their area of application. It is hence always necessary to examine how appropriate particular Standards are to user requirements. This is particularly true in the area of assessment of materials – for instance Federal Test Standard 101C Method 4046!

8.6.1. British Standards Institution

389 Chiswick High Road, London W4 4AL. Tel: +44 181 996 7000 Fax: +44 181 996 7001 email: standards Website: www.bsi. org.uk/bsi

BS 5958: Part 1: 1991 Code of practice for Control of undesirable static electricity: Part 1.
General considerations

BS 5958: Part 2: 1991 Code of practice for Control of undesirable static electricity: Part 2.
Recommendations for particular industrial situations

BS 7506: Part 1: 1995 Methods for Measurements in electrostatics Part 1. Guide to basic electrostatics

BS 7506: Part 2: 1996 Methods for Measurements in electrostatics Part 2. Test methods

8.6.2 European and International Standards: (Available From BSI and other National Standards Organisations, Websites: www.cenorm.be/ and www.cenelec.be/)

CENELEC Report R044-001: 1999 Safety of machinery - Guidance and recommendations for the avoidance of hazards due to static electricity

(Note: this standard is due to be superseded by CLC/TR 50404 Electrostatics - Code of practice for the avoidance of hazards due to static electricity, which is the final satges of development)

EN 1149-1:1995 Protective clothing - Electrostatic properties Part 1. Surface resistivity (test methods and requirements)

(Note: this standard is due to be re-issued as an update during 2003)

EN 1149-2:1997 Protective clothing - Electrostatic properties Part 2. Test method for the measurement of the electrical resistance through a material (vertical resistance)

prEN 1149-3: 2002 Protective clothing – Electrostatic properties Part 3. Test methods for the measurement of charge decay

EN 100015-1:1992 Harmonised system of quality control assessment for electronic components 'Basic specification: Protection of electrostatic sensitive devices Part 1: General requirements'

(Note: this standard is now withdrawn an superseded by the EN version of IEC 61340-5-1)

ISO 6356: 2000 Textile floor coverings - Assessment of static electrical propensity - Walking test

ISO 10965:1998 Textile floor coverings - determination of electrical resistance

8.6.3 International Electrotechnical Commission IEC TC101: (IEC Home Page: Www.Iec.Ch)

IEC TC101 'Dashboard'[Yes, this is all one URL!]: www.iec.ch/cgi-bin/procgi.pl/www/iecwww.p?wwwlang=Eandwwwprog=TCboard.pandcommittee=SCandTC=TC+101

IEC 61340-2-1 Electrostatics – Part 2-1: Measurement methods - Ability of materials and products to dissipate static electric charge

IEC TR 61340-2-2 Electrostatics – Part 2-2: Measurement methods – Measurement of chargeability

IEC 61340-2-3 Electrostatics – Part 2-3: Methods of test for determining the resistance and resistivity of solid planar materials used to avoid electrostatic charge accumulation

IEC 61340-3-1 Electrostatics – Part 3-1: Methods for simulation of electrostatic effects – Human body model (HBM) – Component testing

IEC 61340-3-2 Electrostatics – Part 3-2: Methods for simulation of electrostatic effects – Machine model (MM) – Component testing

IEC 61340-4-1 Electrostatics - Part 4: Standard test methods for specific applications Section 1: Electrostatic behaviour of floor coverings and installed floors

IEC 61340-4-3 Electrostatics – Part 4-3: Standard test methods for specific applications - Footwear

IEC 61340-5-1 TR2 Electrostatics - Part 5-1: Protection of electronic devices from electrostatic phenomena - General requirements

(Note: published in Europe as a full standard under number EN 61340-5-1)

IEC 61340-5-2 TR2 Electrostatics - Part 5-2: Protection of electronic devices from electrostatic phenomena - User guide

8.6.4 US Standards

EOS/ESD Association (7900 Turin Road, Bldg 3, Suite 2, Rome, NY 13440-2069,USA, Tel: +1 315 339 6937, Fax: +1 315 339 6793, email: eosesd@aol.com, website: www.eosesd.org)

The ESD Association has issued a series of Standards and these are best accessed via the ESDA Website

National Fire Protection Association (1 Batterymarch Park, Quincy, Massachusetts 02269-9101

Tel: +1 617 984 7249, Fax: +1 617 770 3500)

Available in Europe from ILI, Index House, Ascot, Berks. SL5 7EU Tel: +44 1344 636 300,

Fax: +44 1344 291 194, email: standards@ili.co.uk, website: www.ili.co.uk

ANSI/NFPA 77 Static Electricity 1993 Edition

NFPA 99 Standard for Health Care Facilities 1996 Edition

8.6.5 Japanese Standards

Japanese Industrial Standard JIS L 1094 - 1988 Testing Methods for Electrostatic Propensity of Woven and Knitted Fabrics

8.7. Shell Safety Guide

'Static Electricity - Technical and Safety Aspects' 1988

8.8. Books

J. A. Cross *'Electrostatics: Principles, problems and applications'* Adam Hilger, Bristol

M. Glor *'Electrostatic hazards in powder handling'* Research Studies Press 1988

W. D. Greason *'Electrostatic discharge in electronics'* Research Studies Press 1992

H. Haas *'Electrostatic hazards: Their evaluation and control'* Verlag Chemie 1977

G. Lüttgens; M. Glor *'Understanding and controlling static electricity'* Expert Verlag 1989

G. Lüttgens; N. Wilson *'Electrostatic hazards'* Butterworth-Heinemann, 1997

D. M. Taylor and P. E. Secker: *'Industrial electrostatics: Fundamentals and measurements'* Research Studies Press, from John Wiley 1994

J. F. Hughes *'Electrostatic particle charging: Industrial and health care applications'* Research Studies Press, from John Wiley 1997

J Pionteck and G Wypych *"Encyclopedia of Polymer and Plastics Additives"* Chemtek Publishing ISBN 1-895198-34-8 First Edition, 2007

Annex 1

DEFINITIONS

The following definitions may help clarify the meaning of a variety of terms used in relation to electrostatics. Several of these definitions are based on BS 7506: Part 1: 1995. Others reflect the views of the author. A start has been made in the IEC to create a list of definitions relating to the microelectronics industry [1].

Antistatic material or treatment: A material or treatment that allows easy migration of charge over its surface or through its volume (i.e. is dissipative) so that when there is a bond linkage to earth there is opportunity for static charge to migrate away to earth. An alternative meaning is that such a material does not easily acquire charge at contact or at rubbing with other surfaces.

Antistatic additive or filler: A substance added to a liquid or solid that makes this an 'antistatic material'

> Note: For plastics this usually involves providing enhanced surface charge dissipation capability by absorption of atmospheric moisture. Performance is then susceptible to ambient humidity.

Bonding: The use of an independent connection between conductors or between a conductor and a dissipative material to provide a path of low electrical impedance for easy migration of charge where this cannot otherwise be ensured

Breakdown: The loss, at least temporarily, of the insulating properties of a medium (gas, liquid or solid) under electrical stress.

> Note: Breakdown in air may be partial, with the occurrence of corona or brush discharge, or complete as in a spark.

Breakdown voltage: The minimum voltage at which breakdown occurs in a given situation.

Capacitance loading: the apparent capacitance exhibited by a material to surface charge divided by that for a thin film of a good dielectric in the same measurement situation and with a similar distribution of charge

> Note: An alternative way of appreciating this is as the surface potential achieved per unit of charge for a thin film of a good dielectric divided by the surface potential achieved per unit of charge with a similar surface charge distribution on the test material.

Charge decay: the migration of charge across the surface or through a material leading to a reduction of surface potential at the area where the charge was deposited

Charge decay time: the time required for the local surface potential to fall to a selected fraction of its initial value. The 'initial value' used may be the initial peak voltage in tribocharging studies or the surface voltage at a similar time after the end of corona charging – typically 100ms. The choice of 100ms is that this is about the minimum time it takes for surfaces to separate in tribocharging actions.

> Note: Convenient decay times for comparison between materials are the time from the initial surface voltage to 1/e (37%) of this (e is the base of the natural logarithm 2.7183) and to 10% of this. Where practical it is desirable to record both times. As the rate of charge decay may vary greatly during the progress of decay it is very useful to record the form of the variation of surface voltage with time.

Conductive material: a material with a high mobility of charge so that any potential on the surface is retained for only a very short time

> Note: The charge decay time of conductive materials is generally much less than 0.05 s.

Conductivity: The reciprocal of resistivity. ($S\ m^{-1}$ equals $(\Omega\ m)^{-1}$)

Conductor: A material providing a sufficiently high conductivity that all parts of it are always at the same potential

Corona: the generation of gas ions of either polarity at a localised region of high electric field and where the field is not sufficiently strong for formation of a spark discharge channel across the gap to other surfaces.

Dissipative material: a material which allows charge to migrate over its surface and/or through its volume in a time that is short compared to the time scale of the actions creating the charge or the time within which this charge will be effective or will cause electrostatic problems.

> Note: For general avoidance of risks and problems in operations involving manual activities the decay time from the initial peak surface potential to 1/e of this needs to be less than 0.25 s and/or within 1s to 10%. To avoid risk of drawing direct sparks from the surface the decay time needs to be greater than 0.01s.

Earth/Earthing: The electrical connection (bonding) of a conductor to the main body of the earth to ensure that it is at earth potential.

Earth bonding point: A dedicated point for connecting people (by a wrist band cord) or equipment to earth.

Electrostatic discharge (ESD): The sudden transfer of electrostatic charge between bodies at different electrostatic potentials by direct electrical contact or by electrical breakdown of an air gap

Electrostatic discharge sensitive device (ESDS): A discrete device, semiconductor, integrated circuit or other assembly that can be damaged by an electrostatic discharge directly to the device or nearby

ESD 'common earth bonding point': A common connection point to which all items in an ESD protected area are connected.

ESD protected area (EPA):. An area in which ESDS can be handled without risk of damage from electrostatic discharges or fields.

ESDS voltage sensitivity: The maximum voltage at which the ESDS does not suffer any ESD damage with a defined type of static discharge

ESDS voltage sensitivity of an assembly: The ESDS voltage sensitivity of most sensitive device in an assembly will determine the sensitivity of the assembly.

Field work: Handling ESDS within a temporary EPA with permanently controlled boundaries.

Flammable material. A gas, vapour, liquid, dust or solid that can react continuously with atmospheric oxygen and that may therefore sustain a fire or explosion when such reaction is initiated by a suitable spark, flame or hot surface. In normal usage 'gas' and 'vapour' are synonymous.

Flammable mixture: A mixture of gases, or of a gas with a mist of liquid or suspension of dust, in which combustion will propagate.

Flammable range: The range of concentrations in air of a flammable material within which combustion can occur.

Garment: A coat, jacket, smock, hood, trousers, overall or cap is regarded as a garment for the purpose of this document

Ground: The uniform potential established in the work area ensuring uniformity of potential of all objects.

Ground cord: An electrical connection between the earth bonding point and the ESD earth facility.

Hazardous area: An area in which flammable or explosive gas-air mixtures are, or may be expected to be, present in quantities such as to require special precautions against ignition.

> Note: Hazardous areas are classified according to whether a flammable mixture is continuously present or only occasionally

Incendive: Capable of igniting a prescribed flammable mixture.

Insulator/insulative material: a material with very low mobility of charge so that charge on the surface is retained there for a long time

> Note: The charge decay time of insulative materials is generally greater than 10 s.

Ion: An atomic or molecular particle carrying electrical charge.

Mass charge density: The nett quantity of charge carried by unit mass of a material.

Materials for packaging: Any material in which ESDS are packed in intimate contact for transportation or storage - including bags, boxes, crates, wraps, magazines, cushioning, foams, loose fill, etc.

Minimum ignition energy: The smallest quantity of energy that can ignite a mixture of a specified flammable material with air or oxygen, measured by a standard procedure.

Non-conductor: see insulator

Relative capacitance: (see capacitance loading)

Relaxation: the migration of charge over and/or through a solid, liquid or gaseous material (see charge decay and dissipation)

Relaxation time: (see charge decay time)

Resistivity: (see surface and volume resistivity)

Surface charge density: The net quantity of charge per unit area of surface of a solid or liquid ($C\ m^{-2}$).

Surface potential: the potential as measured with a non-contacting electrostatic fieldmeter, either with the field sensor close to the surface and its potential adjusted to give zero electric field or measured by the electric field created at the sensing aperture of a fieldmeter at a defined distance and connected to earth.

> Note: It is convenient to express the reading of a nearby non-contacting fieldmeter in terms of the potential on a large plane conducting surface at a defined perpendicular distance. Corrections are needed when the area tested is small. In particular test situations it is appropriate to relate readings to the potential of a conducting surface covering the specific test area of the material

Surface resistivity: The resistance between opposing sides of a square on the surface of the material with account taken of fringing field effects

Triboelectric charging: The separation of charge occurring at separation of contacting surfaces and from relative movement between two surfaces.

Volume charge density: The net quantity of charge per unit volume of a solid, liquid or gas (C m^{-3}).

Volume resistivity: The resistance between opposing sides of a cube of the material with account taken of fringing field effects.

References

[1] *" Electrostatics – Part 1-2: Definitions of all parts of the electrostatics series 61340-x-y"* IEC 61340-1-2 .

Annex 2

TEST METHOD TO ASSESS THE ELECTROSTATIC SUITABILITY OF MATERIALS FOR RETAINED ELECTROSTATIC CHARGE

The following document was prepared by John Chubb as the basis for a new British Standard. This work was carried out in 2004 for British Standards Technical Committee GEL101, which reports to IEC TC101. (Minor corrections and updates to the text were made in 2009 and 2010)

Contents

Foreword

The Test Method described in this Standard provides assessment of the suitability of materials to avoid risks and problems from the influence of static electric charge retained on the surface of a material after contact and rubbing actions with other materials. It is also relevant to the constructive use of charge on materials. The basis of the method is:

a) That most risks from static electric charge retained on materials, and opportunities for constructive applications, relate directly to the local surface voltage created. The surface voltage indicates opportunity for direct air discharge (for voltages over about 300V), opportunity for charge induction on devices nearby (with risks of 'charge device model' type damage events to semiconductors) and opportunity for attraction of airborne particles.

b) That there is a direct relationship between the surface voltage present at the time when practical surfaces separate after contact and rubbing actions and values of charge decay time and capacitance loading shown by the materials. Surface voltages will be limited to low values if the time for decay of surface voltage is very short and/or if the capacitance experienced by surface charge is very high.

c) That the behaviour of materials in practical situations with tribocharging is well represented by measurements on sample areas of the material after corona charging.

1. Scope

The method of test described is concerned with assessing the performance of materials in relation to the influence of static charge retained after contact and rubbing actions. The influence of the surface charge is manifest in terms of the local surface voltage.

This Standard is not concerned with the character of electrostatic discharges to materials or with their shielding capability or with their ability to remove charge from conducting items in contact.

The method of test can be used with any fairly flat surface material (for example as sheet, film and layer) as well as with powders and liquids held in suitable containers.

2. References

2.1. Informative references

A number of published papers relevant to the present Standard are listed in Annex G.

3. Definitions

For the purpose of this standard, the definitions in Annex A will apply.

4. Test Method

Measurements are made on a single layer area of material (for example the fabric of a garment or a sheet or a layer) supported flat from an earthed edge support. Measurements are made both with 'open backing', where there are no earthed or charged surfaces nearby, and with 'earthed backing' where the whole test area rests against a clean flat earthed metal surface. These conditions model the two extremes of practical applications.

The test method is equally applicable to installed surfaces and to powders and to liquids.

A localized patch of charge is deposited in a short period of time in the middle of the test area. This may be deposited by tribocharging or, more conveniently, by a high voltage corona discharge, as described in Annex B. Corona charging may be used as studies have shown that the performance of materials matches well that obtained with tribocharging. A number of relevant references given in Annex G.

The surface voltage created by the deposited charge is measured and the rate of decay of this voltage with time is measured with open and with earthed backing, as described in Annex C.

The quantity of charge transferred by the charging action is measured, as described in Annex D.

The performance of materials is likely to vary considerably with ambient environment of temperature and humidity. These parameters need to be controlled and measured, as described in Annex E.

5. Procedure

5.1. Mounting

Mount the test surface into the corona charge decay apparatus as described in Annex B.

5.2 Environmental conditions

Set conditions of temperature and humidity as described in Annex E. Condition samples in these environments for at least 24 hours

5.3. Pre-test surface voltage

Check pre-test surface voltage is adequately low before each test. This should be less than 2% of the expected or observed initial peak surface voltage achieved at charging. If this voltage is higher than this it is best to wait for it to decay before making a measurement.

5.4. Charge decay times

Measure charge decay times (as described in Annex C) and capacitance loading values (as described in Annex F) using both +ve and –ve polarity charging. Make at least 2 measurements with corona durations of 10 or 20ms at each of the following corona voltages:

+ve (kV)	-ve (kV)
	2.7
3.0	3.0
4.0	4.0
5.0	5.0

NOTE: With this range of corona discharge voltages measure the initial surface voltages and the associated quantities of charge for both polarities over a range of quantities of charge down to a few nanocoulombs (nC). This covers the conditions that relate to tribocharging situations.

NOTE: Additional measurements will be needed if measurements show much variation with position over the tests sample.

5.5 Decay time measurement

Measure the time for the decay of each surface voltage from that at 0.1s after the end of the charging action to 37% (1/e) and to 10% of that at 0.1s.

NOTE: It is not necessary to achieve any specific initial surface voltage, only sufficient voltage for good quality decay time measurement at the end of decay timing.

5.6. Capacitance loading

Calculate values for capacitance loading from each initial voltage value and its associated quantity of charge transferred, as described in Annex F.

5.7. Analysis of capacitance loading values

Plot values of capacitance loading against quantities of charge. Extrapolate the average slopes of variation for positive and negative polarity to zero charge – as illustrated in Annex G. The values of positive and negative capacitance loading for zero charge are noted and averaged.

6. Assessment

The assessment of materials is based on the surface voltage measured 0.10s after the end of a short period of charging.

Judgements are made from either of the two performance features:

a) whether the times for the surface voltage to decay from the initial value observed at 0.1s to 10% of this with open and also with earthed backing is less than a specified time, t(a)

and/or:

b) whether the capacitance loading value extrapolated to zero charge, based on the surface voltage at 0.1s, is greater than N and also that the time for the surface voltage to fall from the 0.1s value to 10% is less than t(b).

For general applications t(a) shall be 1s, N shall be 40, or more, and t(b) shall be 20s.

If it is clear from initial measurements that capacitance loading values are too low for effective control of surface voltages then measurements can concentrate solely on charge decay time measurements.

If it is only practicable to measure the charge decay time (for example with installed surfaces) then this measurement will be sufficient so long as the decay time t(a) is less than the acceptance time, for example 1s.

The 'voltage at 0.10s' is affected by the charge decay after the end of charging and also by the capacitance loading experienced by the surface charge. The choice of 0.1s relates to the time taken in tribocharging actions for the influence of surface charge on nearby items to develop as the contacting surfaces separate. A shorter time will be appropriate for faster separation actions.

If the time for charge decay after the 0.1s is short compared to the time of separation of surfaces, and also if there is a route available for the charge to leak away to earth, then no significant surface voltages can arise.

The maximum surface voltage V_{max} (volts) that may arise in practice for a quantity of charge q (nC) can be obtained from the capacitance loading values extrapolated to zero charge as:

$$V_{max} = f q / (CL_{q=0})$$

- where f is a factor and $CL_{q=0}$ is the value of capacitance loading measured with corona charging extrapolated to zero charge. In practice values for q are likely to be no more than 50nC and the factor f has a value around 75.

NOTE: Extrapolation of capacitance loading values to zero charge has been found the best way to match corona charging to tribocharging performance.

The maximum voltage values derived as above can be compared to risk threshold voltage levels for practical situations. For example, if the maximum surface voltage permissible is 100V then capacitance loading value needs to exceed 40.

In addition to the capacitance loading requirement above, it is necessary that any locally generated charge can drain away to earth. If not then multiple charging actions will lead to a progressive build up of surface potential. Assuming that the material has an earth bonding point (for example the body of a person inhabiting a garment is earthed via footwear or a wrist strap), then the time for charge decay to 10% of the value achieved at 0.10s shall be less than 20s (the t(b) value) with earthed backing. For the case of garments where the area charged may be across a seam from the earth bonding connection then measurements need to be made both where a seam separates the earthing point and the test area and when the earthing connection is on the same area of fabric as that directly charged (as described in Annex F).

7. Test Report

The test report shall include at least the following information:

a) date and time of measurements

b) description and/or identification of material tested

c) corona charging conditions used (e.g. polarity, corona voltages, charging duration)

d) where the sample is supported with an ‘open backing’ and with an ‘earthed backing’

e) individual values of initial surface potential values achieved 0.1s after the end of the charging action and the associated times for this voltage to decay to 37% (1/e) of this and also to 10% of this. Where multiple measurements have been made it is appropriate to plot any variation of charge decay time with quantity of charge. Results shall be shown and quoted separately for measurements with open and earthed backing. Different polarity measurements may be combined.

f) individual values of initial surface potential, quantity of charge transferred and capacitance loading. The variation of capacitance loading with quantity of charge is plotted to show the extrapolation to values at zero charge. Results shall be shown and quoted separately for measurements with open and earthed backing.

Different polarity measurements may be combined.

g) mean charge decay time and the capacitance loading values at zero charge

h) temperature and relative humidity at the time of testing and time for which samples were exposed to these conditions before testing

i) identification of instrumentation used (e.g. type and serial number) and date and certificate details of most recent calibration (as per procedures described in Annex H and Annex J)

ANNEX A: (NORMATIVE) DEFINITIONS

A1 capacitance loading

The surface potential achieved per unit quantity of charge for a thin film of a good dielectric divided by the surface potential achieved per unit of charge with a similar surface charge distribution on the test material

A2 charge decay

The migration of charge across or through a material leading to a reduction of surface potential at the area where the charge was deposited

A3 charge decay time

The time from the initial surface voltage level created by the charge put on to the surface (100%) to a selected, and stated, end point fraction of this. The initial voltage value to be used is that 0.1s after the end of a short period charging action.

> ***NOTE***: *Convenient decay times for comparison between materials are the time from the initial surface voltage to fall to 1/e of this (e is the base of the natural logarithm 2,7183) and to 10% of this.*

NOTE: As the rate of charge decay may vary greatly during the progress of decay it is very useful to record the form of the variation of surface voltage with time.

A4 conductive material

A material with a high mobility of charge so that the potential on the surface is retained for only a very short time

NOTE: The charge decay time of conductive materials is less than 0.05 s.

A5 corona

The generation of ions of either polarity by a high localised electric field

A6 dissipative material

A material which allows charge to migrate over its surface and/or through its volume in a time that is short compared to the time scale of the actions creating the charge or the time within which this charge will be effective or will cause an electrostatic problem.

NOTE: For general avoidance of risks and problems in operations involving manual activities the decay time from the initial surface voltage at 0.1s to 10% of this needs to be less than 1.0 s. To avoid the risk of incendiary sparks the decay time needs to be longer than 0.01s.

NOTE: The dissipative capability of a material does not relate to its ability to remove charge from a conducting item in contact. This ability is determined by resistivity type measurements.

A7 insulative material

A material with very low mobility of charge so that charge on the surface is retained there for a long time

NOTE: The charge decay time of insulative materials is generally greater than 10 s.

A8 relative capacitance

(see capacitance loading)

A9 surface potential

The reading from a non-contacting electrostatic voltmeter or fieldmeter in the test equipment calibrated in terms of the potential on a plane conducting surface covering the equipment test aperture

Annex B: (Normative) Design and Operation of Test Apparatus

B1 Physical design features

A typical arrangement and relevant dimensions of the test apparatus is shown in Figure B1. Other equipment fulfilling the basic design and performance requirements may be used.

The test aperture for deposition and measurement of deposited charge shall be 50 mm ±1 mm diameter or an equivalent quasi-square aperture area. The corona points are mounted on a movable plate in a 10 mm ±1 mm diameter circle, 10 mm ±1 mm above the centre of the test aperture.

The fieldmeter sensing aperture shall be 25 mm ±1 mm above the centre of the test area. When the plate with the corona points is moved fully away, the test area shall be clear up to the plane of the fieldmeter sensing aperture.

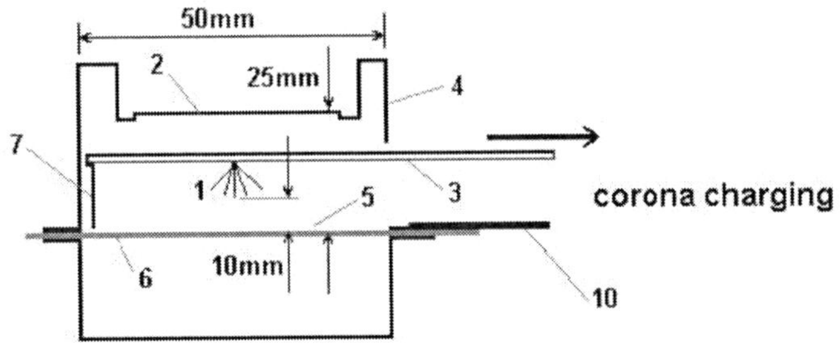

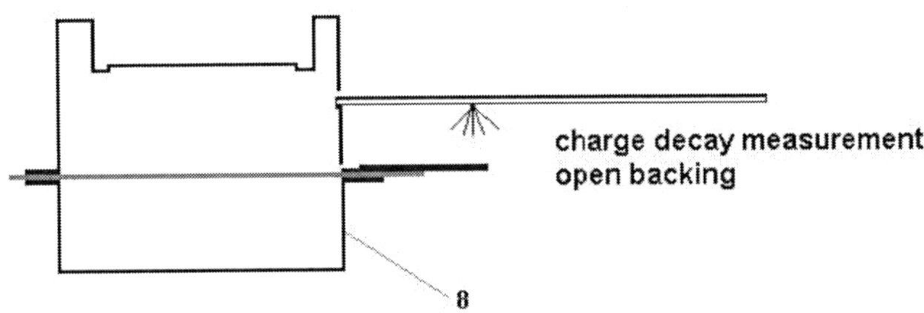

Figure B. Continued on next page.

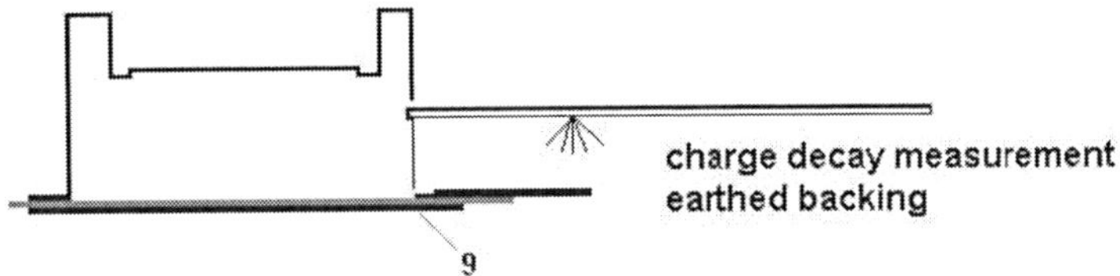

Figure B1 – Example of an arrangement for measurement of corona charge decay and arrangements for 'open backing' and 'earthed backing'

Key to figure 1:

1 - 10 mm diameter circle of corona points

2 - Fieldmeter sensing aperture

3 - Movable plate

- insulating surface mounting corona points
- earthed top surface to shield fieldmeter

4 - Earthed casing

5 - Test aperture: 50 mm -0/+5 mm diameter or (50 mm ± 5 mm) • (50 mm ± 5 mm) area

6 - Specimen under test

7 - Air dam to remove residual corona air ionization

8 - Open shielded backing

9 - Earthed backing

10 - Instrument base plate

B2 Containment of test material

For an installed surface the test aperture in the base plate of the instrument shall be rested directly on the surface.

Sheet or flexible materials shall be tested whilst supported against the test aperture with both 'open backing' and 'earthed backing'. These two arrangements (shown in Figure B1) represent the extreme conditions of practical application. For both arrangements the longer of the two decay times shall be taken for comparison with general acceptance criteria.

In practical terms, 'open backing' measurements represent the condition where materials are well separated from earthed surfaces, for example the bottom edge of a coat or smock hanging away from the body of the wearer. 'Earthed backing' represents the other practical extreme of a material resting in intimate contact with an earthed surface, for example a garment fitted close to the body of the wearer, or resting on a work surface or top of a metal bench. The nature of the material used as an 'earthed backing' surface may affect measurements, so a suitable material needs to be used. For example, the insulating nature of anodised aluminium will inhibit vertical charge migration.

For testing of material with 'open backing' the material shall be supported against the base plate of the instrument by an earthed metal aperture aligned with the instrument test aperture and with at least 5 mm width outside the test aperture area. The shield over the reverse side of the test area shall be earthed and at least 25 mm away over the whole test area.

If charge moves more readily through the bulk test material than across its surface, then decay times may be shorter against an earthed backing. On the other hand, if charge moves more readily across the surface of the test material, then charge decay time may be increased due to increased capacitive loading.

For testing thin flexible films the film needs to be well tensioned to be flat and to reduce risk of flexing during decay time measurements.

Powders and liquids may be tested while supported in an earthed metal cup beneath the test aperture. When testing light powders that may be easily dispersed into the air, precautions need to be taken to avoid powder particles becoming deposited on surfaces around and inside the fieldmeter sensing region. This may be achieved for long decay time materials by, for example, slow retraction of the moving plate and air dam. Alternatvely, by an appropriate increase in the separation between the powder surface and the plane of the sensing aperture.

B3 Corona charge deposition

The size and distribution of charge deposited on the material is not well defined. The arrangement however provides a consistent pattern of deposited charge for decay time and capacitance loading measurement.

The corona charge deposition time shall be 20 ms ± 10 ms. Longer times may be used if needed to achieve an adequate initial peak surface potential for measurement. Corona deposition times longer than 100 ms will not necessarily provide any enhancement of charging and may possibly create some damage to sensitive surfaces. Samples shall be tested with both positive and negative polarity.

> ***NOTE:*** *Typical voltages appropriate for corona charging are between 3kV and 10kV.*

The equipment for charge deposition shall move fully away from the region of fieldmeter observation in less than 30 ms.

> ***NOTE:*** *For corona voltages up to 10 kV the initial surface potential with insulative materials may be up to about 3 kV. For materials with fast charge decay rates and/or high values of capacitance loading the initial surface potential may be much lower - for example only 50 V to 100 V. Corona voltages down to about 3kV may be appropriate for low quantities of charge.*

B4 Fieldmeter

The fieldmeter shall be able to measure the surface potential with an accuracy of ± 5V, or better, with a response time (90% to 20%) less than 10 ms. The stability of the zero shall allow measurement of surface potential with this accuracy over the longest decay times to be measured.

> *NOTE: In measurements of capacitance loading with low corona voltages and small quantities of charge initial surface voltages may be quite low. It will be helpful if surface voltage measurements can be made to an accuracy of ±1V down to zero volts.*

The sensitivity of the fieldmeter shall be set according to the calibration procedure in Annex H to show the surface potential as presented by a plane conducting surface over the full area of the test aperture.

> ***NOTE:*** *A rotating vane 'field mill' type fieldmeter is preferred. Chopper stabilised sensors may be acceptable if the sensitivity, noise levels and zero stability are appropriate. Induction probe instruments are not likely to be suitable, even for fast charge decay measurements, because the influence of even slight residual corona air ionisation will cause zero drift – and the absence of this would need to be tested.*

During corona charge deposition and decay time measurement the sensing aperture of the fieldmeter shall be well shielded from any connections or surfaces associated with corona high voltage supplies. There shall be no insulative materials in or around the region of the instrument between the fieldmeter and the test aperture able to contribute signals to fieldmeter observations.

For measurements with materials having initial peak surface potentials less than 200 V it is necessary to remove residual air ionisation created by the corona discharge when the moving plate carrying the corona discharge points is moved away. An air dam on the trailing edge of the moving plate mounting the corona discharge points is a convenient way to remove this air ionisation from the region between the fieldmeter sensing aperture and the test surface. Residual ionisation shall contribute less than 30 V to measurement of surface potential. This may be tested by measurements on a clean earthed metal test surface.

The value taken of the peak surface potential measured by the fieldmeter will be affected by the initial rate of charge decay and the time for removal of the plate carrying the corona discharge points. When the time for removal of the plate is comparable to the rate of decay, the time for plate movement will affect the value of this peak surface potential. The effect of this is removed by using the voltage at 0.1s after the end of charging as the 'initial voltage' value. Use of the sutfac voltage at this time provides a better comparison to the behaviour ovserved with tribocharging.

ANNEX C: (NORMATIVE) MEASUREMENT OF CHARGE DECAY TIME

C1 Decay time

The 'charge decay time' is the time from an initial voltage created by the charge put on the surface to a selected, and stated, end point fraction of this.

With manual tribocharging there is no influence on items nearby from the charges on the separating surfaces for about 0.1s. This arises from the timescale of body actions and the proximity between the separated charges. The decay of surface voltage immediately after completion of corona charging and removal of the plate carrying the corona electrodes may be of technical interest but it is not relevant to the practical performance and risks that can be created by the surface charge deposited. The initial voltage value to be used is therefore that at 0.1s after the end of the corona charging action.

The fraction of the initial voltage for the end point of charge decay time may be 1/e or 10%. These are designated as $t_{1/e}$ and $t_{10\%}$.

NOTE: The time from the initial voltage to 1/e of this should not be thought of as a 'time constant'. This would imply the decay curve followed an exponential form – and this is generally not true.

C2 Initial voltage

The form of variation of surface voltage with time after the end of the charging action in general varies little with the quantity of charge transferred. Thus the level of voltage reached at 0.1s after the end of the charging action is not directly important. All that is required is that at 10% of this voltage it is still possible to make good quality determination of the end point of timing.

NOTE: Care needs to be taken in the measurement of charge decay times on small signals where signal noise may be significant in relation to signal amplitude. Any signal averaging technique needs to take account of the need for a fast response on the fast initial transients of short decay time curves but then provide suitable averaging for smoothing later.

C3 Voltage decay curve

It is useful to record the form of the charge decay curve. In many cases the rate of decay slows up significantly during the progress of decay and appreciable levels of surface charge may be retained for long periods. This effect is also indicated by comparison of the decay times to 10% with that to 1/e.

NOTE: Records of charge decay curves allow future re-examination of performance

C4 Timing measurement

Decay times may be measured either directly by electronic circuits or on oscilloscope recordings of observations. To overcome signal to noise limitations at low signal levels some local averaging of measurements may be needed - rather than operating simply in terms of maximum and minimum signal values.

C5 Measurement procedure

Where decay times are less than 100 s it is useful to make a number of repeat measurements at the same location and to make measurements with high and low corona voltages and with both polarities of charging. These measurements will demonstrate consistency in material behaviour and any variations of performance with quantity of charge.

NOTE: It is desirable to make measurements to check if corona is causing any change in sample characteristics, both regarding charge decay time and capacitance loading. Any changes by corona may conveniently be examined by making measurements at the same location initially with a low corona charge, then with a high corona charge and then again with a low charge.

It is possiblt that the form of charge decay curves may vary slighty with the quantity of charge deposited. It is wise to make tests over a range of quantities of charge comparable to those likely to arise in the practical situation. Tribocharging by rubbing may involve quantities of charge in the range 10-50nC. It is hence appropriate to make measurements with quantities of corona charge from 50nC downwards.

C6 Overcoming pre-charging

Sample surfaces may become pre-charged by handling when placing ready for testing. It is recommended that when samples are placed in position the moving plate is back, so the fieldmeter can respond to any charge on the sample surface and show these observations. This initial surface potential on relatively insulating samples may be minimised by careful handling with minimum sliding actions.

'Appreciable' pre-charging means surface voltages more than 2% of the expected or measured value of initial peak voltage. Large voltages than this may affect the results of decay time measurements to 10%.

Two main options are available if there is appreciable pre-charge:

- to wait until the pre-charge has dissipated. This means waiting for the initial surface potential to fall to below the 2% level

- to make a study on the decay of this pre-charge without adding any corona charge. This means making a measurement with the corona voltage turned off or set for zero volts. It is to be noted that the decay of such pre-charge may be rather slower than the decay of the local patch of corona charge. It is none the less a useful observation.

It is not recommended that quality measurements are attempted by putting corona charge on to an already well charged surface or material.

It is also not recommended that pre-charge on the material is neutralised by any means other than waiting. Deposition of neutralising charge may only give the appearance of neutrality by creating close-coupled regions of charge.

C7 Additional artefacts to avoid

There are three other possible artefacts are worth noting:

- that if the air dam on the leading edge of the moving plate touches the sample surface then tribocharging may occur. This may arise when testing light fabrics. The fabric surface should be stretched flat under the test aperture, but it may still rise by induced air movement. This effect may be checked by making measurements with no corona charging. It may be avoided by slightly raising the base plate of the instrument off the sample.

- that with some materials a very short (1 ms to 2 ms) transient peak surface potential may be observed before the real charge decay curve. This is usually positive. Occurrence of this transient may upset operation of software timing. This is thought to be due to vertical charge separation between front and back surfaces of the sample at sample flexing.

- that if static charge is retained on the surface of the moving plate facing the test surface then charge may be drawn over the sample by the induction electric field. This effect may be limited by minimising the non-earthed area on the underside of the moving plate.

C8 Calibration

The performance of the decay time measuring instrumentation shall be assessed according to the calibration procedure in Annex H.

ANNEX D: (NORMATIVE) MEASUREMENT OF THE QUANTITY OF CHARGE TRANSFERRED WITH CORONA CHARGING

The charge received by a sample surface with corona charging can be measured using a sample support arrangement shown in Figure D1.

The aperture for the sample in the mounting plates needs to be larger than the test aperture in the base plate of the charge decay test unit to avoid direct corona charge flow to these plates. This may be tested by checking for negligible 'conduction' charge signal in the absence of a sample. The performance of the surface charge measuring instrumentation shall be assessed according to the calibration procedure in Annex J.

The corona charge transferred to the test surface is measured in two parts:

1) as charge directly coupled to the sample mounting plates

2) as charge remaining where it is deposited.

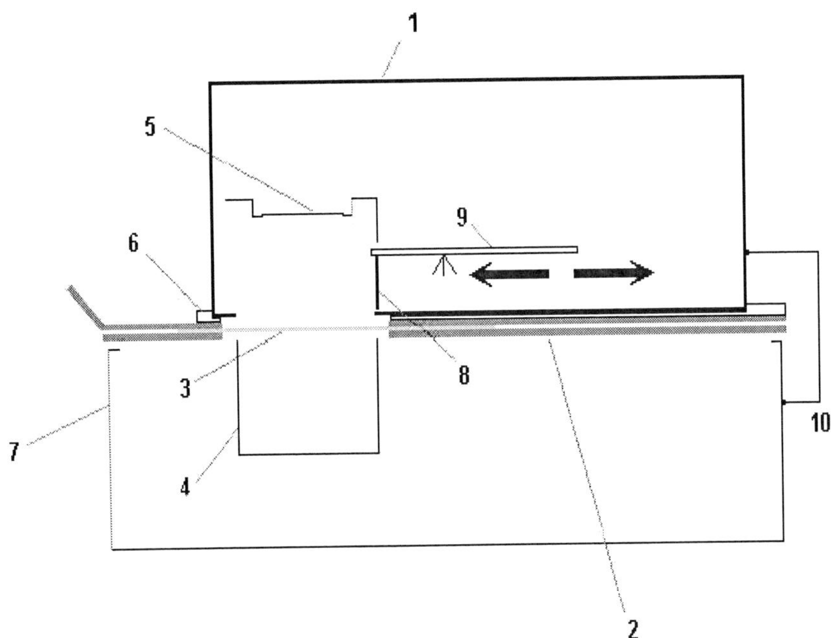

Figure D.1 – Arrangement for measuring received charge

Key:
1 Charge decay test unit
2 'Conduction' charge measuring support plates
3 Sample
4 'Induction' sensing electrode
5 Fieldmeter sensing aperture
6 Insulation between charge decay test unit and sample mounting plates
7 Shielding box
8 Air dam
9 Moving plate with cluster of corona discharge points
10 Earth bonding link from charge decay test unit to shielding box and 0V of charge measurement circuits

The 'conduction' charge couples directly, and within the time of the observations, to the mounting plates and is measured by a suitable virtual earth charge measurement circuit. The charge remaining in the area of deposition is sensed by an 'induction' electrode beneath the open-backed sample. If the form of the induction sensing electrode is similar to the mechanical form of the charge decay apparatus above the test area then the retained charge will couple about half to the equipment above and half to the induction electrode below. The induction charge component will then be about twice the charge received by the induction electrode.

> *NOTE: The apparently simple approach of measuring the charge leaving the charge decay test unit will not give correct values for charge. The reason is that if charge is retained on the surface of the test material, and does not move quickly out to the sample mounting plates, it will couple back to the structure of the unit and so not be fully available for measurement. Equally, charge transfer cannot be measured from that leaving the corona electrodes as part of this charge may flow directly to the structure of the instrument.*

The precise sensitivity of induction charge measurements and overall calibration of the charge measurement unit is determined as specified in Annex J and H.

ANNEX E: (NORMATIVE) ENVIRONMENTAL CONDITIONS

E.1 Standard environmental conditions

The charge decay properties of materials usually vary with temperature and relative humidity. It is hence very important that the values of these parameters in the test environment are known, and that they are controlled to agreed values for standardised measurements.

To assess whether materials are significantly susceptible measurements should be made with at least 2 levels of humidity, one around 50%RH and the other below 25%RH. If materials are to be used in conditions outside this range of the above standard values (in particular at lower levels of humidity) then their characteristics must be checked under the most extreme conditions likely to be encountered and these conditions reported.

For standardised laboratory measurements samples shall be conditioned in the selected environmental conditions for at least 24 h before testing and measurements made with the conditions shown in Table 1 and in the listed sequence, or as otherwise agreed.

Table E.1 – Environments for conditioning and testing

		Temperature °C	**Relative humidity** %
	High relative humidity	23 ± 2	50 ± 3
	Low relative humidity	23 ± 2	12 ± 3

For measurements in practical situations the ambient temperature and relative humidity shall be recorded.

Samples need to be stored in the set environmental conditions in ways that allow air circulation around both surfaces for at least 24 hours.

Operation of the test equipment may generate a little heat. This will change the temperature and humidity conditions to which the sample is exposed inside the equipment relative to those outside and to which the samples have accommodated. Precautions need to be taken to reduce such changes.

NOTE: It is desirable to measure temperature and humidity within the test equipment to establish that there are no problems. It is also desirable to store samples with easy air circulation and not in the test position.

E.2 Sample handling and preparation

Handle samples only well away from the area of the sample to be tested or use tweezers. Avoid breathing in the direction of the sample.

The surface of the material tested shall be clean and free of loose dust. Remove any loose dust by gentle brushing or blowing with clean, dry air. If the surface is obviously contaminated either an alternative area or sample shall be tested, or measurements made with the contamination present and the condition of testing reported to be 'as received'.

NOTE: Solvent or chemical cleaning is NOT recommended, as this may change surface conditions.

For measurements in practical or installed applications, the materials shall be tested without any "special" cleaning. If cleaning is part of the process, for example washing of garments, measurements should be taken before and after cleaning where practical. The materials and the method used to clean shall be reported.

ANNEX F: (NORMATIVE) ASSESSMENT OF CAPACITANCE LOADING MEASUREMENTS

F.1 Calculation of 'capacitance loading' values

The capacitance effect experienced by charge on the test surface is best defined as the ratio of the surface potential achieved per unit of charge for a thin film of a good dielectric divided by the surface potential achieved per unit of charge with a similar surface charge distribution on the test material. This may be thought of as a 'capacitance loading' or as the 'relative capacitance'.

The capacitance loading experienced by charge on the surface is obtained from measurement of the charge received by the test surface, Q, and the initial surface potential, V, observed. The value is calculated by comparing the observed ratio of initial surface potential per unit of charge to the ratio observed with a very thin layer of good dielectric (e.g. cling film) that has a sufficiently short decay time that a low pre-test surface potential can be achieved in a reasonable time. This ratio is equivalent to the ratio of the apparent capacitance value calculated for the test material C to that for a very thin layer of good dielectric, C^*. The following equations shall be used to calculate capacitance loading:

The apparent capacitance of reference material (very thin layer of good dielectric) is:

$$C^* = Q_{ref} / V_{ref} \qquad (1)$$

Apparent capacitance of test material:

$$C = Q / V \qquad (2)$$

Capacitance loading:

$$CL = C / C^*$$

$$CL = (Q / V) / (Q_{ref} / V_{ref}) \qquad (3)$$

- where Q_{ref} is the total charge received and V_{ref} the initial surface potential observed on the reference material, Q is the total charge received by the test material, and V is the initial surface potential observed on the test material.

Once a value has been obtained for the apparent capacitance $C^* = Q_{ref} / V_{ref}$ for the reference material in the particular test arrangement this may be used as a reference value in subsequent capacitance loading measurements, so long as all features of the test arrangement remain the same. This should be checked from time to time.

F.2 Interpretation of capacitance loading measurements

With some materials (for instance cleanroom garment fabrics) the capacitance loading varies with the quantity of charge deposited. In such cases the values of capacitance loading calculated from each measurement of initial voltage at 0.1s after the end of corona charging and the associated quantity of charge are plotted against the quantities of charge involved – as illustrated in Figure F1 below. The average slope of the linear variations for positive and negative polarity are extrapolated to zero charge.

Extrapolation of capacitance loading values to zero charge has been found the best way to match corona charging to tribocharging performance.

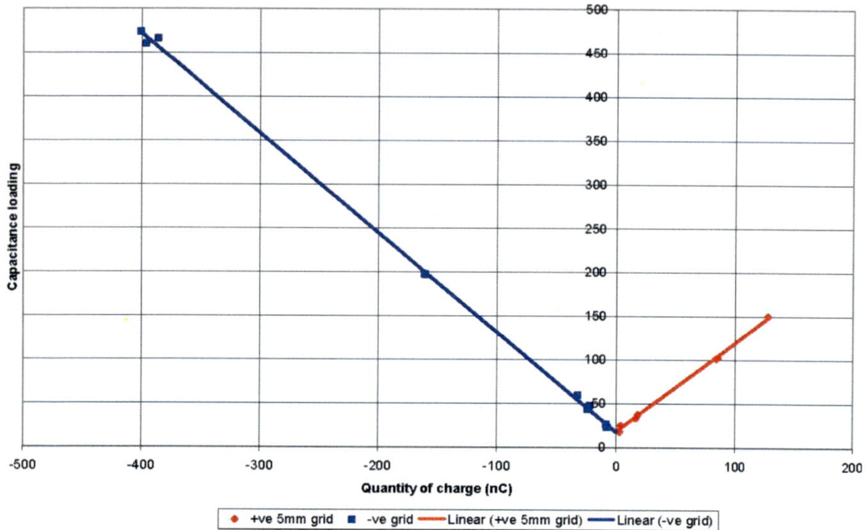

Figure F.1: Example of variation of capacitance loading with quantity of corona charge

ANNEX G: (INFORMATIVE) BIBLIOGRAPHY

[1] R. Gompf *"Standard test method for evaluating triboelectric charge generation and decay"* NASA Report MMA-1985-79 Rev 2, July 1988

[2] J. N. Chubb *"Instrumentation and standards for testing static control materials"* IEEE Trans Ind. Appl. 26 (6) Nov/Dec 1990 p1182.

[3] J. N. Chubb and P. Malinverni *"Experimental comparison of methods of charge decay measurements for a variety of materials"* EOS/ESD Symposium 1992 p5A.5.1

[4] J. N. Chubb *"Dependence of charge decay characteristics on charging parameters"* 'Electrostatics 1995', York April 3-5, 1995 Inst Phys Confr Series 143 p103

[5] J. N. Chubb *"Corona charging of practical materials for charge decay measurements"* J. Electrostatics 37 1996 p53

[6] J. N. Chubb *"The assessment of materials by tribo and corona charging and charge decay measurements"* 'Electrostatics 1999' Univ Cambridge, March 1999 Inst Phys Int Confr

[7] R. Gompf, P. Holdstock, J. N. Chubb *"Electrostatic test methods compared"* EOS/ESD Symposium, Sept 26-30 1999

[8] J. N. Chubb *"Measurement of tribo and corona charging features of materials for assessment of risks from static electricity"* IEEE Trans Ind. Appl. 36 (6) Nov/Dec 2000 p1515.

[9] J. N. Chubb "New approaches for electrostatic testing of materials" *J. Electrostatics* 54 March 2002 p233 (*Proceedings ESA2000 Annual meeting,* Brock University, Niagara Falls, June 18-21, 2000)

[10] J. N. Chubb, P. Holdstock, M. Dyer *"Can cleanroom garments create electrostatic risks?"* 'Cleanroom Technology' 8 (3) March 2002 p38

[11] J. N. Chubb, P. Holdstock, M. Dyer *"Can surface voltages on inhabited garments be predicted"* Inst Phys 'Electrostatics 2003' Conference, IoP Conference Series 178 2004 p131

[12] J. N. Chubb, P. Holdstock, M. Dyer *"Predicting maximum surface voltages on inhabited cleanroom garments in practical use"* ESTECH Phoenix 18-21 May 2003

[13] P. Holdstock, M J D Dyer, J N Chubb *"Test procedure for predicting surface voltages on inhabited garments"* EOS/ESD Symposium 2003. Las Vegas. 21-25 Sept 2003

[14] J. N. Chubb *"Comments of methods for charge decay measurement"* J. Electrostatics 62 2004 p73-80.

[15] J. N. Chubb *"A Standard proposed for assessing the electrostatic suitability of materials"* J. Electrostatics 65 2007 p607-610

[16] J N Chubb, H Walmsley *"Charge decay time prediction"* To be published in J Electrostatics in 2010

[17] J N Chubb, J Harbour *"Signal processing for charge decay measurement"* Submitted for publication in Brit J Appl Phys E. February 2010

Annex H: (Normative) Calibration of Corona Charge Decay Measuring Instrumentation

H1 Aspects to be calibrated

Calibration of charge decay measurement instrumentation involves two parts:

1) calibration of the surface potential sensitivity of the fieldmeter

2) calibration of the decay time measurement performance.

H2 Equipment

Calibration of the charge decay measuring instrumentation is made using a plane conducting surface covering the whole test aperture area with a small separation (less than 0.5 mm) below the edge of the test aperture so calibration voltages can be applied. A suitable arrangement is shown in Figure H1 below.

H3 Surface potential sensitivity calibration

The surface potential sensitivity calibration is made in terms of a uniform potential on a conducting surface covering the whole test aperture area.

The voltage source shall provide a stable, low ripple voltage of both polarities to at least ±1000 V. The voltage measuring system shall cover the measurement of both polarities and be separate from the voltage source so it may be formally calibrated independently. The accuracy of voltage measurement shall be better than 0.2%. The stability of the calibration voltage shall be 0.2%.

H4 Decay time calibration

Calibrated resistors and capacitors are connected in parallel between earth and the conducting calibration plate over the test aperture. The resistors and capacitors shall be of good quality, with linear characteristics with voltage and be capable of withstanding voltages up to 3 kV.

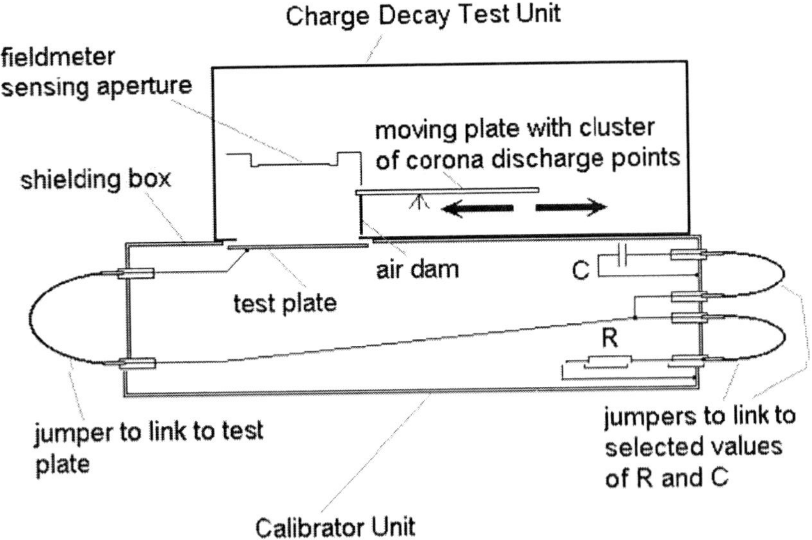

Figure H1: Arrangement for charge decay instrument calibration

Decay time values in seconds are derived from the product of the values of the resistors (ohms) and capacitance (farads). Decay time values shall be provided for each decade of time over the main operating range of the instrument. To cover the range of interest of materials used for static control the decay time values provided should cover the range 0.1 s to 100 s.

Formal calibration of the resistors and capacitors shall be made in the equipment as connected and as used for calibration of the charge decay instrumentation.

H5 Calibration procedure

The charge decay time measuring instrument is mounted on the calibration equipment, switched on and allowed to stabilise. Connect the calibration plate to earth and measure the initial 'zero' surface potential reading by the fieldmeter. Apply calibrated voltages to the plate to give readings at well spaced voltage levels from 50 V to 1000 V. Repeat measurements for the other voltage polarity.

Connect a combination set of resistance and capacitance values from earth to the calibration plate. Operate the charge decay measuring instrument to apply sufficient charge to the calibration plate to achieve an initial peak surface potential suitable for decay time measurement. Initial surface potentials in the range 100 V to 1000 V are convenient. Measure the time from the initial surface potential to 1/e of this using the normal instrument charge decay time measurement facilities. If both electronic and software decay time measurement facilities are available then both shall be used together.

At least 3 decay time measurements shall be made for each charge polarity for each decay time value setting. From each set of 6 readings the average decay time value and the standard deviation shall be calculated and recorded.

H6 Calibration Certificate information

The calibration certificate shall include to following information:

a) Name of orgaisation issuing the certificate

b) Certificate number

c) Customer name

d) Instrument type number

e) Instrument serial number

f) Date of calibration

g) Name and signature of authorised signatory

h) Method of calibration and instrumentation used

i) Assessment of overall accuracy

j) Reference information on date and source of calibration of measuring instruments used and accuracy of their calibration

k) Table of results as:

- list of applied voltages, positive and negative
- readings obtained: table of decay times with standard deviations and values from combinations of calibrated resistors and capacitors

ANNEX J: (NORMATIVE) CALIBRATION OF CORONA CHARGE TRANSFER MEASURING INSTRUMENTATION

J.1 Aspects to be calibrated

Calibration of instrumentation for measurement of the corona charge received by samples involves two parts:

1) calibration of the charge sensitivity of the 'induction' and 'conduction' charge measurement circuits

2) calibration of the interpretation of induction charge observations.

J.2 Induction and Conduction charge measurement sensitivity

A defined quantity of charge can be provided in two ways:

- by charging a calibrated capacitor to a defined voltage and discharging this directly to the electrodes involved in 'conduction' and in 'induction' charge measurement. For 'virtual earth' charge measurement circuits all the charge on the capacitor is reliably transferred to the charge measurement circuit.

- for 'virtual earth input' charge measurement circuits a defined quantity of charge can be provided by a defined flow of current for a defined time. The current is determined from a defined and calibrated voltage source and a defined and calibrated value of series resistance with the current flow switched between earth and the 'virtual earth nout' circuit to minimise capacitive effects.

The signal output is compared to the known quantity of charge input.

J.3 Relative sensitivity of 'induction' charge measurement

The relative sensitivity of 'induction' charge observations is the charge signal measured compared to the quantity of charge placed on the sample surface at the position of charge deposition. This is best determined by making charge decay studies with the charge decay test unit mounted on the charge measuring sample support using a test sample that is a thin and fairly homogeneous dielectric and has a charge decay time of several seconds. Charge measurement signals will initially be solely a 'induction' charge signal. This will progressively move to become just a 'conduction' charge signal. Since the total charge is constant the relative sensitivity factor is the factor by which the decaying 'induction' signal needs to be multiplied so that when this is added to the increasing 'conduction' signal the sum, that is the total charge value, is stable over the time of observation.

The variations of conduction Q_c and induction Q_I signals are best recorded numerically over the initial period of charge decay, for example to the 1/e time. Spreadsheet modelling may then be used to find the numerical factor, f_I, by which instantaneous 'induction' signals need to be multiplied so that when added to corresponding instantaneous 'conduction' signals a total signal Q_{tot} is achieved that does not vary over the observation time.

$$Q_{tot} = Q_c + f_I * Q_I \qquad (1)$$

The process is illustrated in Figure J2.

Once a value has been obtained for the factor f_I this may be used as a reference value in subsequent charge measurements, so long as all features of the test arrangement remain the same.

J4 Calibration Certificate information

The calibration certificate shall include to following information:

a) name of orgaisation issuing the certificate

b) Certificate number

c) Customer name

d) Instrument type number

e) Instrument serial number

f) Date of calibration

g) Name and signature of authorised signatory

h) Method of calibration used

i) Assessment of overall accuracy

j) Reference information on date and source of calibration of measuring instruments used and accuracy of their calibration

k) Table of results as a list of applied quantities of charge with associated readings for positive and negative negative polarities for both the induction and the conduction electrodes

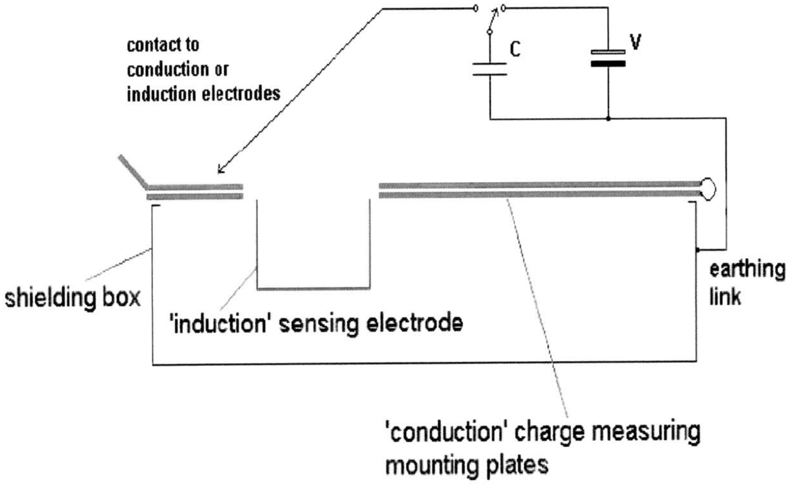

Figure J1: Arrangement for calibration of charge sensitivity using a charged capacitor

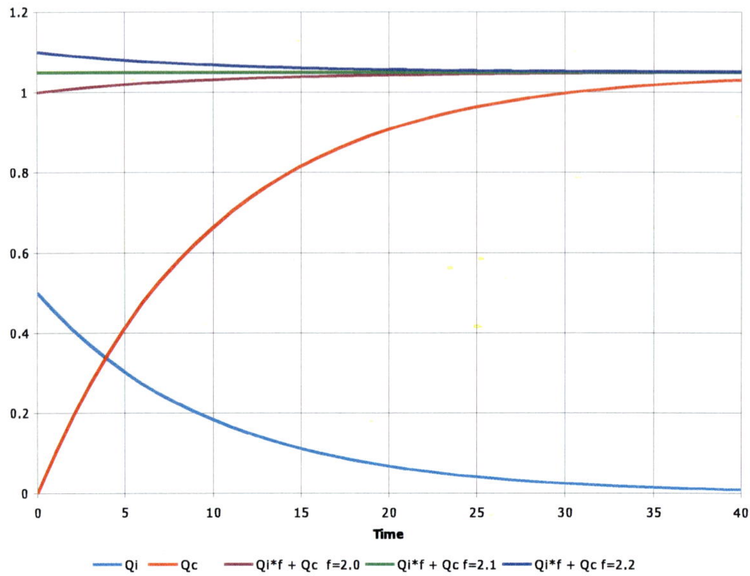

Figure J2: Adjustment of factor f for matching induction to conduction measurements

Annex 3

METHODS FOR THE CALIBRATION OF ELECTROSTATIC MEASURING INSTRUMENTS

1. INTRODUCTION

Methods are described in the following sections for formal calibration of the main instruments for electrostatic measurements. They cover:

- electrostatic fieldmeters for measurement of electric field, surface voltage, space voltage and surface charge density
- electrostatic voltmeters for surface voltage and zero current drain voltage measurements
- Faraday Pails for measurement of charge
- charge decay test units for assessment of the suitability of materials
- air ionization test units

The methods described are based on the methods described in British Standard BS 7506: Part 2: 1996 [1] but have been updated in the light of experience. No comparable alternative methods are available in Standards literature. Methods for the calibration of instruments for the measurement of voltage, resistance, capacitance and separation distance, etc are available in the Standards literature. There are many Test Houses available to provide direct formal calibration of these parameters to National Standards.

2. Common Features

2.1 Introduction

All instruments used in calibration work should be calibrated with reference to national or international standards according to the procedures and requirements of BS EN 30012-1. The following sections describe a number of common requirements for calibration of electrostatic measuring instruments.

2.2. Voltage Source

Voltages are typically required of both polarities from ±50V up to ±30kV. The source needs to be stable and with low ripple to better than 1% and preferably to 0.2% of the voltage level applied. More than one power supply unit may be needed to cover the required range with adequate stability and ease of adjustment.

Voltage levels for calibration should be used from less than 25% of the most sensitive full scale range to at least 25% of the least sensitive range. They should cover 100% of all ranges except the least sensitive.

2.3. Voltage Measurement System

The voltage measuring system should cover the full range of measurement needed for both polarities. It should preferably be separate from the voltage source so that it may be calibrated and used independently. Voltage measurements should be made with direct independent connection to the calibrator plates or instrument and the accuracy should be better than 0.2% for high accuracy and better than 1 % for medium accuracy instruments.

Voltages up to 1000V may be measured using a digital multimeter. For higher voltages a high voltage resistive divider should be used to present a known fraction of the voltage to a precision digital voltmeter.

For voltages and voltage dividers working at over 1000V precautions should be taken to avoid corona as corona discharge currents could affect accuracy. To avoid the input impedance of the measuring voltmeter influencing accuracy, the high voltage divider should be calibrated in conjunction with its measurement voltmeter. Resistor values of 1000M and 1M are convenient for the high and low voltage arms of a divider.

The voltage values at which the voltage measuring instruments are calibrated should be those at which calibration measurements are made to avoid any linearity errors at interpolation between calibration values.

2.4 Resistors and Capacitors

High voltage capacitors and resistors are used for the calibration of charge decay instruments and charge plate monitors. These should be calibrated in situ in the unit or set-up used for calibration. Calibration of the values of capacitance and resistance should be made to better than 1%.

2.5 Distance Measurements

Distance measurements should be made using slip gauges. Calibration of these should be checked every two years.

2.6 Temperature and Humidity

Sensors used to provide supporting temperature and humidity information should be formally calibrated. If incorporated into instrumentation they should at least be precalibrated.

2.7. Standard Instrument Calibration Certificate Information

The calibration certificate needs to include the following information:

a) the name of the organization issuing the certificate
b) certificate number
c) customer identity
d) instrument description and type number
e) instrument serial number
f) date of calibration

g) name of person who carried out the calibration and name and signature of authorized signatory
h) identification of method of calibration used (e.g. reference number of Standard)
i) overall assessed accuracy of each aspect of calibration
j) physical measurement information relevant to the calibration set-up
k) reference information on the date and place of calibration of measuring instruments used and the accuracy of these calibrations
l) list of applied parameters with derived values and the actual instrument readings observed with values for upper and lower ranges where sensitivity ranges overlap

The calibration certificate of an instrument dispatched to a customer direct from manufacture should record the 'as dispatched' values. The calibration certificate of an instrument returned from a customer should record the 'as received' calibration, to the extent that the condition allows this, as well as the ‘as dispatched’ values. If no adjustments are needed the calibration values presented on the calibration certificate are designated 'as received and dispatched'. If it has been necessary to change components or make adjustments then the instrument needs to be recalibrated in the sealed condition ready for dispatch and both the 'as received' and 'as dispatched' values are presented on the calibration certificate.

An adhesive label should be attached to the instrume nt to show date and source of calibration.

3. Electrostatic Fieldmeters

3.1. Apparatus

The electric field for calibrating electrostatic fieldmeter instruments is set up by application of a stable continuous voltage between a pair of plane and parallel metal plates [1,2]. The fieldmeter instrument is mounted with its sensing aperture flush with the inner surface in the middle of one of the plates (see Figure 1). The size and spacing of the plates is such that the electric field in the central region is not affected by fringing field effects at the plate boundaries or by the presence of any charged surfaces outside the plate system. The spacing of the plates is such that perturbations of the field between the plates, by the sensing aperture structure of the fieldmeter, do not extend across to the further plate.

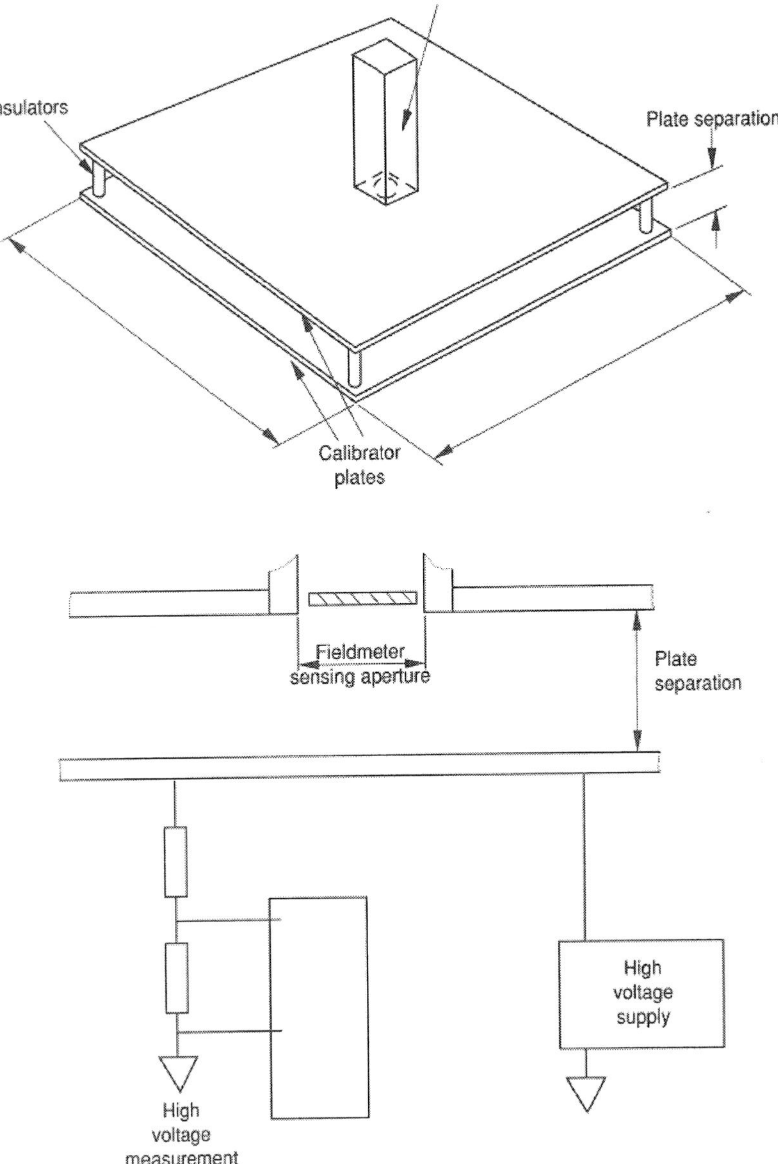

Figure 1.General arrangement for calibration of electrostatic fieldmeters.

For calibration to within 1%:

- the spacing between the plates should be at least 1.5 times the sensing aperture diameter;
- the radial extent of the plates should be at least 15 sensing aperture diameters.

For calibration to within 5%:

- the spacing between the plates should be at least 1.5 times the sensing aperture diameter;
- the radial distance should be at least 9 diameters.

Adequacy in the radial size of the calibration plates against external electric fields may be tested by shorting the plates together and observing any signal change on the most sensitive range when a piece of charged plastic is brought near the outside of the plate gap.

The mounting hole should be a close fit for the spigot around the sensing aperture. The fit should be better than 0.5% of the sensing aperture diameter and with the sensing aperture coplanar with the surrounding surface to within ±0.1% of this diameter. The error in matching the plane of the sensing aperture to the plane of surrounding inner surface of the mounting calibration plate is established using slip gauges to measure the thickness of the mounting plate at four equispaced points around the mounting hole and measuring the height of the spigot of the fieldmeter calibrated.

The calibration plates should be rigid, flat to better than 2 % of plate spacing, smooth and free from contamination and loose dust. The rigidity should be adequate to avoid any change in plate separation by the loading of the heaviest fieldmeter instruments to be calibrated. If in doubt, the spacing between the plates should be measured with a fieldmeter load present.

The outer edges of the plates should have radii of curvature of 2 mm or more and/or be covered by a local layer of insulation to avoid corona discharges at the higher calibration voltages (for example over 5kV).

If separation of the plates is achieved by stand-off insulators between the plates these should be mounted at the outer periphery of the plates so that any charge trapped on the insulators during high voltage operation has no influence of the electric field in the central region. This is particularly important when the plates are shorted together to check the zero setting of the fieldmeter.

With the calibration system in the 'as used' condition the spacing distance between the plates is measured using slip gauges. Measurements of separation distance are taken at four equispaced positions close around the fieldmeter mounting hole. The separation spacing is taken as the arithmetic mean of measurements with calculation of the uncertainty.

3.2. Procedure

Calibration is achieved by comparing the fieldmeter reading with the value of electric field provided by dividing the voltage applied by the separation gap: the measurements are repeated over a range of applied voltages.

Mount the fieldmeter on the calibration unit with its sensing aperture flush with the inner surface it its mounting plate. Switch on the high voltage supply and allow it to stabilize.

With the calibrator plates shorted together take the initial (zero) reading of the fieldmeter display and/or output signal.

Apply voltages between the calibrator plates to give readings from around 25% of the most sensitive range up to at least 25% of the least sensitive range of the fieldmeter. For autoranging instruments use voltages which give readings less than 25% and more than 90% of each range so there is some overlap between ranges. Use the voltage values for calibration at which the voltage measuring system has been calibrated.

Repeat the calibration measurements for both polarities.

Note 1. Calibration to 90% of full scale may not be feasible where the least sensitive range is over 500 kV m^{-1}. After increasing the calibration voltage to maximum, check the readings as the voltage is decreased with a specific check on the zero reading.

Note 2. Differences between increasing and decreasing values of voltage or changes in zero reading may be due to charging of any insulation in the sensing region of the fieldmeter or to dust between the calibration plates. It may be necessary to clean both the upper and lower calibrator plates.

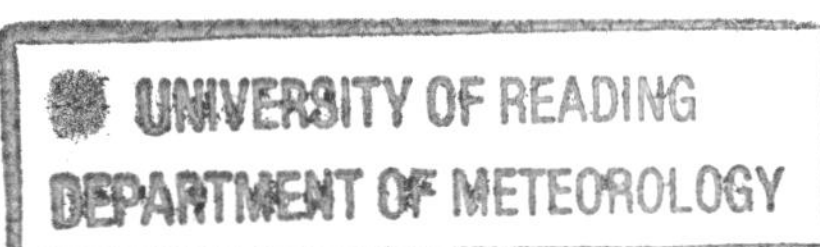

3.3. Results

The results of calibration are presented in a Table of instrument readings with corresponding the values of voltages applied and calculated values of the electric field for both polarities.

3.4 Common aspects of calibration

Details of calibration measurements and associated information should be recorded as given in Section 2.7.

3.5 Calibration Certificate Information

The calibration certificate needs to include the following in addition to the standard information listed in 2.7. a Table listing values of applied voltages, the corresponding values of electric field and the actual instrument readings observed for both polarities (with values for both upper and lower ranges where sensitivity ranges overlap).

4. Proximity Voltmeters

4.1. Apparatus

Proximity electrostatic voltmeters come in two basic forms:

4.1.1 Electrostatic Fieldmeter:

The electric field observed at the sensing aperture of an earthed electrostatic fieldmeter at a defined separation distance from the surface to be measured relates directly to the voltage on the surface. Readings are not linearly dependent on gap and can be influenced by any electrostatic charges in the vicinity.

4.1.2. Voltage Follower Probe:

A voltage follower probe is a fieldmeter mounted close to the test surface so that the electric field readings are determined only by the nearby surface and there is no influence from any electrostatic charges in the vicinity. The voltage needed to be applied to give zero electric field is then the surface voltage

4.2. Arrangements for Calibration

4.2.1 Electrostatic Fieldmeter:

Proximity voltmeters are calibrated by mounting the instrument with its sensing aperture at the specified 'normal' operating distance perpendicular to the middle of a large plane metal surface and recording the reading of the instrument as a function of voltage applied to the metal plate. The calibration arrangement is shown in Figure 2. For calibration to ±1 % accuracy, the radial extent of the calibrator plate should be at least five times the separation distance between the sensing aperture and the plate. There should be no surfaces nearer than 1 m which can retain static charge and no earthed surfaces nearer than 0.5 m. Insulators used to mount or support the calibration plate should be on the opposite side to the voltmeter.

Note: Many proximity voltmeters are set for an operating distance of 100 mm so a radial extent of five times the separation distance requires a calibration plate of at least 1 m square.

The calibration plate should be smooth, free from contamination and loose dust and flat to better than + 2 % of the voltmeter separation distance. Plate edges should have a radius of curvature of 2 mm or more and/or be covered by a local layer of insulation to avoid corona discharges at the higher calibration voltages.

The spacing between the fieldmeter sensing aperture and the plate for earthed fieldmeter instruments should be measured using slip gauges. The distance shall be measured with an accuracy better than 0.5 % for high accuracy instrument and 2 % for medium accuracy. All surfaces that could become charged and influence readings need to be kept well away from the sensing unit – at least ½m.

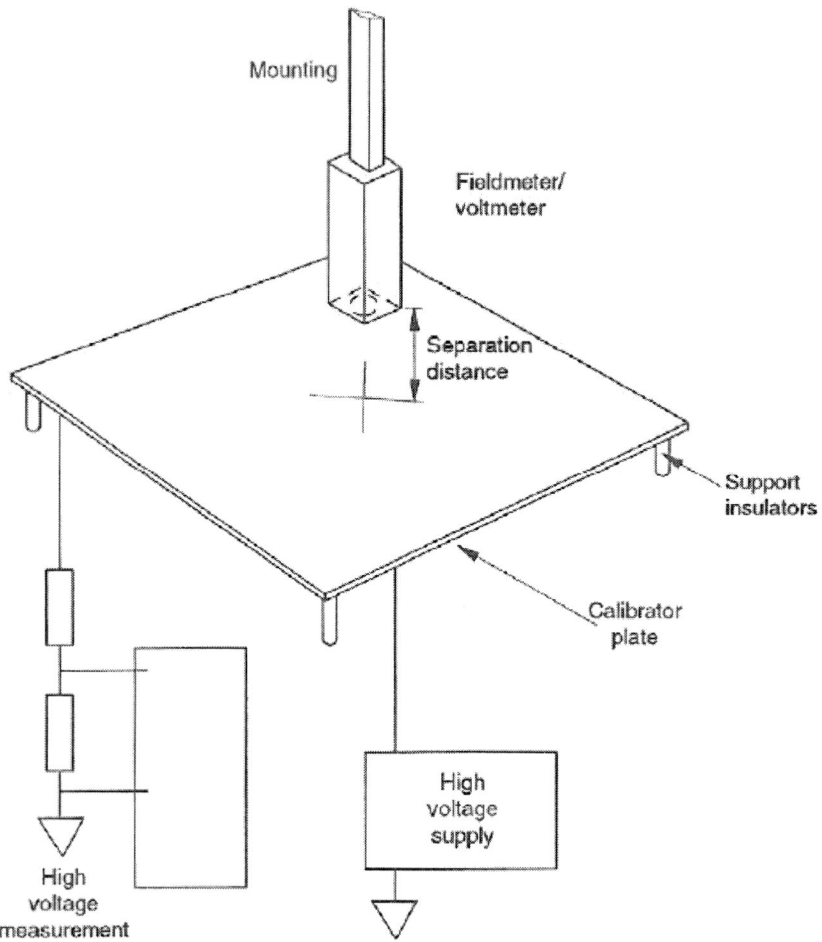

Figure 2. general arrangement for calibration of a fieldmeter proximity voltmeter.

4.2.2 Voltage Follower Probe:

The probe head unit should be mounted by an insulating support with the sensing aperture separated from the clean flat calibration plate a distance similar to the size of the sensing aperture. The mounting insulation must be suitable to withstand the highest calibration voltage to be used. It must also be well shielded to avoid any possibility that residual charge can influence observations. The separation distance is not measured so long as it is less than

10 % of the minimum radius radial distance of the surface surrounding the sensing aperture.

4.3 Voltage Source And Measuring System

The voltage source and voltage measurement system should fulfill the requirements described in Sections 2.2 and 2.3.

4.4. Procedure

Mount the voltmeter at the specified separation distance, connect it to earth, switch on and allow it to stabilize. Read the initial (zero) value with the calibration plate shorted to earth.

Apply voltages to the plate to give readings from around 25 % of the most sensitive range up to at least 25 % of the least sensitive range. For multirange instruments use voltages which give readings from less than 25 % to more than 90 % of each range. Readings should be made on both ranges where there is overlap between ranges.

Measure the output for both increasing and decreasing voltages including zero for both polarities.

Note 1. Differences in readings between increasing and decreasing calibration voltage may be due to charging of insulation in the sensing region of the voltmeter, dust on the calibration plate or to charge on nearby surfaces.

Note 2. The influence of any initial charge on nearby surfaces may be tested by checking the zero readings of the voltmeter when mounted to a clean metal 'zero check chamber' and then as mounted into the calibration position with the calibration plate connected to earth.

4.5. Common Aspects of Calibration

Details of calibration measurements and associated information should be recorded as given in Section 2.

4.6. Calibration Certificate Information

The calibration certificate needs to include the following in addition to the standard information listed in 2.7: a Table listing values of applied voltages, with corresponding values of instrument readings observed for both polarities (with values for both the upper and lower ranges where sensitivity ranges overlap).

5. Electrostatic Voltmeter

5.l. Calibration method

Calibration is achieved by applying calibrated voltages to the input terminal and recording the readings against the applied voltages. If the voltage measurement unit is separable from the Electrostatic voltmeter system then calibration must include this identified unit.

5.2. Voltage Source and Measuring System

The voltage source and voltage measurement system should fulfill the requirements described in Sections 2.2 and 2.3.

5.3 Arrangements for Calibration

If the measuring region of the voltmeter is fully earth shielded from the surrounding environment there is no need to control charges on nearby surfaces. If the voltmeter assembly is open and readings are susceptible to charges nearby then care must be taken to earth shield the surroundings and also the operator.

5.4 Procedure

Switch on the voltmeter and allow it to stabilize. Earth the input connection and record the zero reading. Apply voltages to give readings from around 25 % of the most sensitive range up to at least 25 % of the least sensitive range. For multi-range instruments use voltages which give readings less than 25 % and more than 90 % of each range so there is some overlap. Repeat the measurements for both polarities.

5.5. Results

List the values of applied calibration voltages and the corresponding instrument readings for both polarities.

5.6. Common Aspects of Calibration

Details of calibration measurements and associated information should be recorded as given in Section 2.7.

5.7. Calibration Certificate Information

The calibration certificate needs to include the following in addition to the standard information listed in 2.7:

A Table listing the readings observed for each value of applied calibration voltage for both polarities (with values for both the upper and lower ranges where sensitivity ranges overlap).

6. Faraday Pail

6.1. Calibration Method

The charge introduced into a Faraday Pail is best measured directly using a virtual earth charge measuring amplifier. The approach of measuring charge by the increase in the voltage of the capacitance of the pail system may be convenient, but because it operates by charge sharing between the item or material introduced and the capacitance of the pail there will be an associated, but perhaps small, adjustment needed. With the virtual earth measurement the pail remains at earth potential so all the charge received into the pail is transferred to the measurement circuit and a more precise measurement is made.

There are two approaches to calibation that may be used: discharge of a charged capacitor and a timed period of defined current flow.

6.1.1 Charged capacitor

The principle with the charged capacitor is to charge a calibrated capacitor to a calibrated voltage and discharge this into the input of the charge measurement circuit. The virtual earth measurement circuit ensures the input connection remains at earth potential and all the charge from the charged capacitor ($Q = CV$) is transferred to the feedback capacitor with the voltage appearing as the output voltage.

It is necessary to use a good quality capacitor to ensure linearity of capacitance value with applied voltage and avoid any influence of charge leakage from the time of charging to the time of connection to the charge measurement circuit. The test voltage should not be less than 10V to minimize the risk of influence from contact potentials. It is wise to discharge the capacitor via a resistor to limit the maximum inrush current to within the current drive capability of the virtual earth input amplifier stage. A resistor of 10,000 ohms is likely to be suitable. The capacitance value as calibrated needs to include that of any leads or connection links involved.

6.1.2 Defined current flow

A defined and stable flow of current can be achieved from a calibated reference voltage source and a calibrated precision resistor. The flow of this current is initially to earth and is electronically switched to flow into the input of the virtual earth measurement circuit for a defined period of time. As the current flow is continuous there is no influence from any distributed capacitance in the precision resistor or its connections. The electronic switch and the layout of the circuit need to be chosen that gives negligible charge injection at operation. The period for current flow can be derived with high accuracy and stability by scaling down from a quartz crystal oscillator. Each of the 3 factors determining the quantity of charge output (voltage, resistance and time) are to be formally calibrated with reference to National Standards.

6.2. Voltage Source and Measuring System

The voltage source and voltage measurement system should fulfil the requirements described in Sections 2.2 and 2.3.

6.3 Arrangements for Calibration

Calibration of Faraday Pail systems based on a virtual earth charge measurement circuit are essentially immune to the capacitance of the pail and any linking cable capacitance. It may none the less be wise to make formal calibration of the complete system to confirm this.

Calibration with charge measurement by the increase in pail voltage must take place with the earthed shield around the pail in place. Absence of charge on nearby surfaces should be checked and precautions taken to avoid influence by choice of operator clothing and shielding surfaces nearby. Where low voltages are used for calibration a check needs to be made for the possible influence of electrochemical potential differences between materials. A simple check is to observe readings with zero voltage applied to the contact and use this as an offset for subsequent readings. For high sensitivity measurements it may be necessary to gold plated contacting surfaces to avoid electrochemical voltage effects.

6.4. Procedure

6.4.1. Method 6.1.1:

For direct charge sensitivity calibration: switch on the Faraday Pail charge measurement circuit and allow it to stabilize. Zero the display or output signal reading by shorting the feedback capacitor in the virtual earth circuit, release and then monitor readings for zero drift over a time comparable to normal measurement time.

For charged capacitor calibration: connect one pole of the calibration capacitor to the earth connection of the Faraday Pail system. The other pole of the capacitor is then connected first to the calibration voltage source and then moved to contact the pail. Apply quantities of charge to the pail to give readings from around 10 % of the most sensitive range up to at least 25 % and preferably up to 95% of the least sensitive range. It is desirable there is some overlap across multiple ranges. Repeat the measurements for both polarities.

To avoid test voltages less than 10V with multi-range instruments it may be necessary to use a number of values of capacitor. The quantity of charge transferred by a capacitor C (Farads) at a voltage V (volts) is Q = CV coulombs.

For the timed current flow method of calibration: connect the calibrator to the input of the charge measurement circuit or to the Faraday Pail and allow time for any residual charge on cable connections to dissipate – checked by observing stability of zero reading. Apply quantities of charge to the pail to give readings from around 10 % of the most sensitive range up to at least 25 % and preferably up to 95% of the least sensitive range. It is desirable there is some overlap across multiple ranges. Repeat the measurements for both polarities.

6.4.2. Method 6.1.2:

Measurement of the capacitance of the pail in the Faraday Pail system requires care if the capacitance value is low – for example less than 100pF. Measurements need to take place with the shield in place around the pail. It is necessary to take account of lead capacitance and how this may vary as contact is made to the pail. The simplest approach is to have the measuring contact supported on a rod of good insulation so the lead connecting to the capacitance meter is not close to the hand. The pail capacitance can be measured as the change in capacitance value as the measuring connection is moved from just a short distance away from contact and into touch contact with the pail.

For voltage sensitivity calibration: switch on the Faraday Pail voltage sensor system and allow it to stabilize. Earth and then isolate the Faraday pail from earth and take readings over a time comparable to normal measurement to check for zero reading drift. Apply voltages of both polarity to the pail to give readings from around 10 % of the most sensitive range up to at least 25 % and preferably up to 95% of the least sensitive range. It is desirable there is some overlap across multiple ranges. Repeat the measurements for both polarities.

The charge sensitivity is calculated from the capacitance of the pail C (Farads) and the voltage sensitivity V (volts) as Q = CV coulombs.

6.5. Results

The average value and the standard deviation should be calculated for each set of calibration measurements.

6.6. Common Aspects of Calibration

Details of calibration measurements and associated information should be recorded as given in Section 2.7.

6.7. Calibration Certificate Information

The calibration certificate needs to include the following in addition to the standard information listed in 2.7: A Table listing the averaged readings observed for each set of values of voltages and capacitances used for both polarities (with values for upper and lower ranges where sensitivity ranges overlap).

7. Charge Decay Time Measuring Apparatus

See Annex 2: Annex H: (Normative) Calibration of corona charge decay measuring instrumentation.

See Annex 2: Annex J: (Normative) Calibration of corona charge transfer measuring instrumentation

8. Air Ionisation Test Units

Air ionization test units are calibrated in two parts:

- measuring the voltage sensitivity by applying defined voltages of both polarity to the test plate
- measuring the time for the plate voltage to decay from a set initial voltage to a set end point percentage of the initial voltage. For air

ionization units the initial voltage is chosen as ±1000V and the end point voltage as 10% of this – 100V.

Measuring instrument requirements and calibration procedures follow those for charge decay test units – Section 7. It needs to be noted that due allowance must be made for the capacitance of the test plate assembly to the values of formally calibrated capacitors used in conjunction with calibrated resistors to create defined decay time events.

REFERENCES

[1] *"Methods for measurements in Electrostatics"* British Standard BS7506: Part 2: 1996

[2] J N Chubb *"The calibration of electrostatic fieldmeters and the interpretation of their observations"* Inst Phys Conference 'Electrostatics 1987' Inst Phys Confr Series 85 p261

Acknowledgments

I would like to acknowledge my debt to all the people who have worked with me over the years for their help, encouragement and support. I would particularly like to thank Ian Pollard, Len Gowland and Herbert Watson for their help while I was at Culham Laboratory and John Harbour, Peter Folland, John Miles and Katy Hand for their efforts in making the business of John Chubb Instrumentation Ltd a success. I also wish to acknowledge my debt to the UKAEA Culham Laboratory, where I worked 1962 to 1978, for the opportunities to investigate topics in depth and in an atmosphere of searching after truth without need for expediency. Finally, I wish to acknowledge all the help and encouragement given to me over the years by my wife, Patricia. Without her support this book would not have been written.

Cheltenham 2010

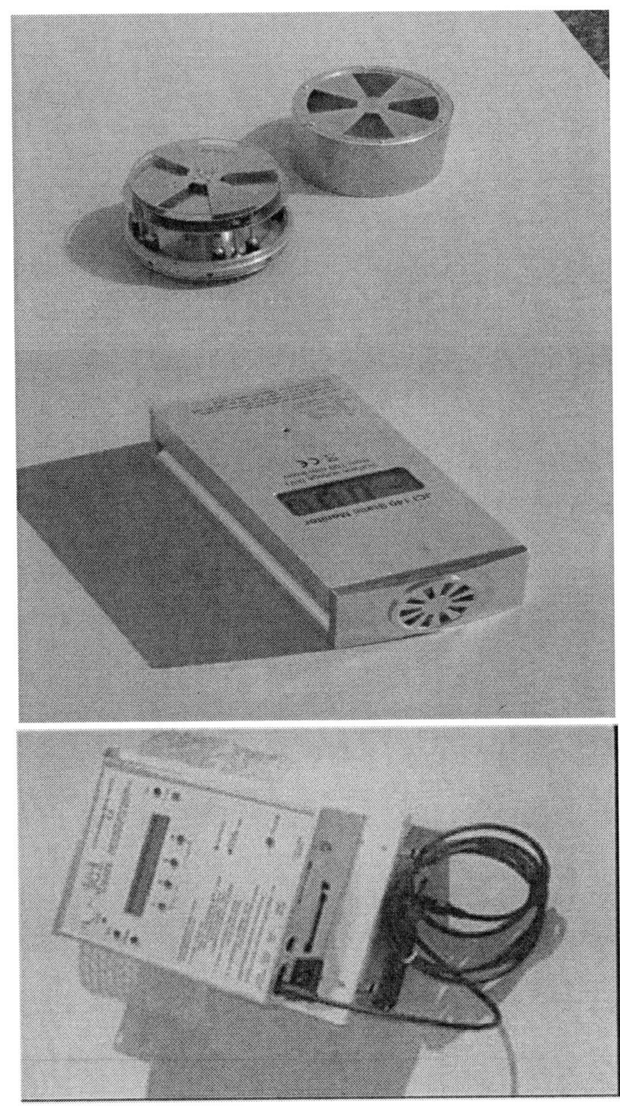

INDEX

A

B

C

D

E

F

G

H

I

Q

R

S

T

U

V

W